T0215817

BestMasters

Mit „BestMasters" zeichnet Springer die besten Masterarbeiten aus, die an renommierten Hochschulen in Deutschland, Österreich und der Schweiz entstanden sind. Die mit Höchstnote ausgezeichneten Arbeiten wurden durch Gutachter zur Veröffentlichung empfohlen und behandeln aktuelle Themen aus unterschiedlichen Fachgebieten der Naturwissenschaften, Psychologie, Technik und Wirtschaftswissenschaften.

Die Reihe wendet sich an Praktiker und Wissenschaftler gleichermaßen und soll insbesondere auch Nachwuchswissenschaftlern Orientierung geben.

Georg Zimmermann

From Basic Survival Analytic Theory to a Non-Standard Application

 Springer Spektrum

Georg Zimmermann
Paris-Lodron University of Salzburg
Austria

BestMasters
ISBN 978-3-658-17718-8   ISBN 978-3-658-17719-5   (eBook)
DOI 10.1007/978-3-658-17719-5

Library of Congress Control Number: 2017937111

Springer Spektrum
© Springer Fachmedien Wiesbaden GmbH 2017

Printed on acid-free paper

This Springer Spektrum imprint is published by Springer Nature
The registered company is Springer Fachmedien Wiesbaden GmbH
The registered company address is: Abraham-Lincoln-Str. 46, 65189 Wiesbaden, Germany

Foreword

The whole story began on Nov 20, 2013, when I received an email from Professor Arne Bathke. He asked me and a colleague of mine if we were interested in analyzing an epilepsy dataset. Since the medical researchers involved in that project had got stuck due to difficulties concerning the statistical methods, they sought for help from mathematicians. To cut it short - some details can be found in the epilogue - I agreed, and so, finally, my master's thesis evolved from this medical research question.

My master's thesis is organised as follows: At first, I want to introduce the reader to some basic research questions and problems in order to point out the need for appropriate survival analytic methods. In addition to that, I give a brief description of the dataset mentioned above, which is used to illustrate the theoretical concepts throughout my master's thesis.

The introductory chapter is followed by some definitions and statements concerning essential quantities one has to be familiar with when working on survival analysis. In this context, I also discuss a basic nonparametric approach, namely the Kaplan-Meier and the Nelson-Aalen estimators.

Next, the focus is on regression models for survival data. Due to the fact that many different models are available, only two of the most popular approaches, the Cox model and the Weibull model, are discussed in detail. When applying the concepts of survival analysis to real-life data, special attention must be paid to the question whether the basic assumptions of the method chosen by the investigator are fulfilled or not. Therefore, I discuss several means of model checking as well as some problems which may arise in applications.

The final chapter deals with the main research question the medical experts sought an answer for: What can we tell about the life expectancy of a patient with certain characteristics? For a start, I introduce measures in the survival analytic setting and in the context of life table calculations which are useful for answering the question stated above. I continue with discussing how to make inference about those quantities. Moreover, it is emphasized that one faces several problems when seeking for a proper methodology to calculate life expectancies.

At the end of my master's thesis, I review the main results of the previous chapters by giving an outline of the "pathway" from getting some real-life dataset to presenting the results of its analysis. Thus, I hope that the interested reader gets a more vivid idea of what applying statistics means.

Most parts of my master's thesis are based on the textbooks of Klein and Moeschberger [16] as well as Kalbfleisch and Prentice [14]. Moreover, I have used further publications on specific topics. All calculations have been carried out using the statistical software package R (R Core Team [22]).

Going back to the initial email from Professor Bathke mentioned above, I think that he was absolutely right when he said that the medical research team "made a very favourable impression". So, I would like to thank Dr. Yvonne Höller, Dr. Claudia Granbichler and Professor Eugen Trinka for treating me, being an undergraduate, with so much respect and trust. Moreover, I am deeply grateful that Professor Arne Bathke has encouraged me in my work very much. Finally, I owe special thanks to my family, especially to my parents, who have supported me all the way.

With respect to this monograph, I would like to thank Carina Berg from Springer Verlag for considering my master's thesis for publication. Moreover, I am grateful to Britta Göhrisch-Radmacher from the editorial office of Springer. She supported me throughout the whole editing process of this monograph.

Salzburg, February 2017   Georg Zimmermann

Contents

1 Introduction

Survival analysis, or, more generally speaking, the analysis of time-to-event data, is an important issue in mathematical theory as well as in many fields of applied statistics. To begin with, at least some basic examples concerning waiting times are done at school. When studying Mathematics at university, the discussion of "waiting time distributions" like the geometric or exponential distribution is part of every introductory course to Stochastics. If you are interested in probability theory, you maybe take a class on Markov chains, where you also learn something about waiting times. One of the reasons why you should at least know some basic facts about this type of mathematical problems is that waiting times are an important issue in a broad variety of applications: For instance, you can study the radioactive decay as well as the reliability of machine components, the mean waiting time at the ski lift in a certain skiing area or the number of games until your first win at the roulette table.

However, the terms "time-to-event analysis" and "survival analysis" usually refer to a certain bulk of methods and do not comprise everything that has to do with waiting times. As the reader will learn throughout this text, survival analysis makes extensive use of concepts which are well known from other statistical contexts, e.g. regression modelling. Another example would be the use of Markov chains in the so-called competing risk setting, but this topic won't be covered in my master's thesis. What all the different methods of survival analysis have in common is that they allow for a certain phenomenon called *censoring*. This feature can be rightfully considered as the key point in survival analysis.

"Censoring" means that the survival time of at least one individual is not observed "completely" (of course, I will give an exact definition later). There can be various reasons for censoring, which is best illustrated by taking a look at some examples. First, consider a (hypothetical) cancer study where the survival times of 1000 cancer patients were recorded from Jan 1, 2004 until Dec 31, 2013. The collected survival data may have one or both of the following features:

- At the end of the follow-up period, 100 patients were still alive. In other words, they experienced the event of interest **after the end of the study period**. Consequently, the survival times of those persons must be at least 10 years.

- Moreover, 10 patients were **lost to follow-up during the study period**, e.g. because after Jan 1, 2005, they didn't come to the annual check in the hospital any more. All we know about those patients is that their survival times must be at least one year.

What the two possibilities mentioned above have in common is the fact that we don't know the exact survival times of those patients. We only have partial information about their survival experience insofar as we know that their survival times are at least as long as the follow-up times of the patients.

Let's now turn to another (hypothetical) example, this time from the technical sciences. A company specialized in manufacturing electronic components wants to analyze the times to failure (i.e., the "survival times") of their products. Let's assume a study period of six months. Furthermore, a certain code containing the production date is assigned to each component, so that the failure times can be easily calculated. In contrast to the example from the medical field above, we can be quite sure that none of the components would get lost to follow-up because if the customer detects a defect, he would moreorless immediately call the company to report the failure. But of course, it can happen that, let's say, 7,000 out of 10,000 components which were produced in total are still working at the end of the six-months period. So, again, we don't know the exact "survival times" for all components. The only thing we know about those 7,000 components is that they were doing their job for at least the time span between production and the end of the study period.

Next, we shall see that there are also other forms of censoring which have not been discussed so far: For example, consider a study in which the subjects were asked, "At which age did you first drink alcohol?" (a similar example can be found in Klein and Moeschberger [16], p. 17). Although some may remember the moreorless exact age, it wouldn't be suprising if some couldn't recall when the first time had been. Consequently, if "age of first use" is considered as time to event, we only know that in some cases, the event of interest has occured prior to the current age of the subject. In contrast to the examples above, where we knew that the time to event exceeds the observation time, we can only say here that the time to event is less or equal to the observation time (i.e., the current age). According to these relations between the exact time-to-event and the observation time, the first example demonstrates *right censoring*, whereas the latter one shows *left censoring*.

By the way, it might well be possible that there are also some people who give the answer "I never drank alcohol." This would obviously represent a case of right censoring. Thus, you see that one and the same study can comprise different types of censoring.

Another censoring scheme which arises quite naturally in particular from the settings of studies in the medical field is called *interval censoring*. As the terminus technicus suggests, this form of censoring occurs, for instance, when patients are scheduled for periodic follow-up after the surgical excision of a tumor. Consequently, if the event of interest is recurrence of a tumor, the exact event times can only be observed if they fall into one of those "observational windows".

For now, this brief, example-driven outline of different censoring types should do. The key message is that there can be many reasons for the occurence of censoring. I hope the examples mentioned above made clear that this is not just an abstract mathematical problem which is hardly seen in applications. Of course, one might ask why not to circumvent those difficulties associated, for example, with right censoring by lengthening the follow-up time. In the medical area, this way of solving the problem is often not possible due to the large amounts of money it would cost. However, in our example concerning the production of electronic components, that may well be feasible: The way of getting the data is quite easy and not expensive at all since if a component fails, the customer will definitely call the company to report the failure and ask for repair. However, the company may not be willing to take the time to wait until all components produced in a certain period (e.g., a year) eventually fail: If it takes, let's say, three years until knowing the exact failure times of each component, it would be definitely too long if the main interest of the study was, for example, to check whether some machines involved in the production process do a good job or not. A similar argument can be applied to medical issues: Usually, one of the primary goals in a cancer survival study is to get some reliable information the doctors can tell the patients. Consequently, it would not make sense if you had to wait for, let's say, twenty years until all the patients of your initial study cohort are dead (maybe, the clinician who brought the study on the way would have retired in the meantime). Moreover, we should be aware of the fact that, depending on the topic, large study times can lead to further methodological problems, as will be demonstrated by real-life data later (see chapter 5).

So, all in all, we see that there isn't a proper way to get around the problem of censoring. Thus, we really need to find appropriate methods which allow to take censored data into account. This issue will be covered in detail in the next chapter.

Now, we turn to the description of a real-life dataset which will be used to illustrate the theoretical concepts throughout my master's thesis. This dataset contains measurements of a huge amount of dates, demographic and clinical factors from 4280 epilepsy patients who were examined at the outpatient clinic for seizure disorders at the Department of Neurology, Innsbruck, Austria, between Jan 1, 1970 and Dec

31, 2010 (this data has been partially reported in Granbichler et al.[10] and Trinka et al.[27]). For sake of simplicity, I only mention those variables which are used in the examples throughout the text. To begin with, we have to accomplish the important task of defining the starting point and the event of interest. As to the first one, we make a quite common choice and take the date of first visit at the clinic, which is referred to as *behandlungsbeginn* in the database. Next, we define "death" as event of interest. So, we, literally speaking, conduct a survival analysis. The information on the date of death is contained in a column named *todesdatum*. Note that these dates were collected by using probabilistic record linkage methodology (see Oberaigner and Stühlinger [19]). Therefore, censoring only occurs if the patient is still alive at the end of the study period since as far as the survival times are concerned, patients lost to follow up are not an issue here. Now, we are ready to calculate the survival time t by taking the time span between *behandlungsbeginn* and either *todesdatum* or Dec 31, 2010, where we choose the latter date if and only if there's no value for *todesdatum* available (i.e. the person has not died yet and thus represents a censored observation). Accordingly, we define a censoring indicator named s which is set to 0 if the observation is censored and 1 otherwise. Note that the terminus "censoring indicator" may be misleading since one usually expects that with respect to the question whether a observation is censored or not, 1 means "yes". But all standard textbooks dealing with survival analysis define it the other way round, and therefore, I decided for this definition, too. By the way, the censoring scheme underlying the data will be more thoroughly examined in the next chapter.

Before we can take a first (descriptive) look at the survival times, we have to discuss one more issue. In our dataset, there's a column named *anfallsbeginn* which contains the date of seizure onset. For some patients, the time span between this date and the date of diagnosis is fairly large, i.e. several years. But, from a medical point of view, the onset of seizures can be in some way regarded as the onset of epilepsy. So, if we are interested in the time from epilepsy onset to death, we

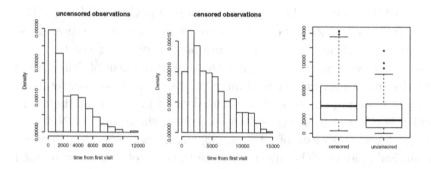

Figure 1.1: Histograms and boxplot of the uncensored and censored observations in the epilepsy dataset.

for sure get biased results due to the discrepancies between the date of seizure onset and the date of diagnosis (remember that we use the latter variable as starting point!). To circumvent this problem, we only take those subjects for which the difference between the two dates mentioned above is less than or equal to one year. Furthermore, we delete a few observations which obviously contain typesetting errors (e.g., the date of seizure onset exceeds the date of diagnosis, which doesn't make sense for medical reasons). Finally, we thus get a subset of the original data consisting of 1217 observations. This dataset is the one I will use throughout the following chapters, unless stated otherwise. To begin with, let's have a look at the uncensored and censored survival times separately.

Example 1.1. In the dataset, there are 250 uncensored and 967 censored observations. The histograms as well as the boxplots for the observed survival times and the censoring times (see Figure 1.1) indicate that both distributions are highly skewed, with the majority of the probability mass belonging to relatively small values of survival and censoring time, respectively. The boxplots show that there is a bit more variability in the censored times, as indicated by the fact that both the range and the distance between the first and third quartile are larger than the corresponding values for the uncensored times. The median of the censored times is 3837 days (about 10.5 years), whereas the median of the uncensored times is 1863 days (approximately 5 years). Note that in this case, it's obviously a good idea to take the median as descriptive location measure, since the plots show skewed distributions and the presence of some outliers.

Of course, however, these quantities must be interpreted with caution: They are only very overall measures of the data because we haven't accounted for any covariates such as age, gender etc. At least for the censored times, we get some interesting information: In our setting, right censoring means that the subjects are still alive at the end of the study period. Therefore, the value of the median tells us that about one half of the censored observations were diagnosed after 2000 (recall that we don't have any patients lost to follow up!). Thus, using the information about the absolute subject numbers from above, we know that more than one third of the entire patient cohort was diagnosed, let's say, in the last 10 years of the study period.

Next, we examine some of the variables recorded in the database. For sake of simplicity, we only take a look at those variables which will be included in the survival analytic models discussed in the following chapters. Throughout the text, these covariates are referred to as *gender, age at diagnosis* (i.e., the age at first visit at the clinic), *year of diagnosis* (i.e., the year of the first visit at the clinic) and *epilepsy etiology* (symptomatic, idiopathic and cryptogenic).

Example 1.2. To start with, we have data of 684 men (145 uncensored, 539 censored) and 533 women (105 uncensored, 428 censored) recorded.
Next, let's examine the age at diagnosis of the patients. When looking at the histogram (see Figure 1.2), we see a maybe multimodal distribution with two "local maxima" between, roughly speaking, 20 and 30 and around 60, respectively. The ages at diagnosis range

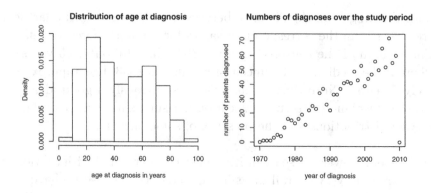

Figure 1.2: A histogram for the variable *age at diagnosis* and a plot showing the numbers of diagnoses in a certain year.

from about 0.4 to 99.4 years, with a median age at diagnosis of approximately 42.2 years.

As to the three main epilepsy etiologies, we state that the majority of the patients, namely 724 subjects, have symptomatic epilepsy, followed by 303 patients with cryptogenic and, at last, 190 patients with idiopatic epilepsy.

Finally, let's have a look at the number of diagnoses in a certain year, that is, the number of patients entering the study in one of the years from 1970 to 2010 (Figure 1.2). Obviously, there is a linearly increasing trend, which in particular agrees with our previous finding that a substantial amount of the study cohort was diagnosed in the 2000's.

For the analyses carried out in the following chapters, it is instructive to look at the sizes of several subgroups formed according to different covariate combinations. For now, we restrict ourselves to patients with uncensored survival times (the reason for this will be clear when looking at the Kaplan Meier curves in the next chapter). Doing so, we see that in some groups, there are hardly any or even no patients at all.

For example, we only have two men with idiopatic epilepsy who were younger than 25 at time of diagnosis. If we change the restriction on *age at diagnosis* to > 65, we haven't got any entries in our database at all! Similarly, we only have one woman with symptomatic epilepsy who was younger than 40 at time of diagnosis. So, the key message is that although we have a relatively large number of uncensored observations (250), we may face problems concerning estimation at least in some subgroups of the data. This issue will be discussed again in the next chapter.

2 Basic terminology and quantities

We first describe how survival time can be appropriately modeled and define some basic quantities which characterize the survival time distribution. Then, we introduce some terminology and notation of survival time samples. In this context, we will also precisely state what censoring means. At the end of this chapter, we will discuss how some of the basic quantities introduced before can be estimated. Throughout this chapter, we will follow the textbooks of Klein and Moeschberger [16], pp. 21-36, 63-72, 91-101 as well as Kalbfleisch and Prentice [14], pp. 6-18.

Basically, we interpret an observed survival time, denoted by t, as a realization of a nonnegative random variable T. We assume T to be continuous, although it should be mentioned that there are also survival analytic methods available for discrete survival time as well as for mixtures of those two types. For more details, we refer to standard textbooks such as Kalbfleisch and Prentice [14], pp. 8-10.

Recall that the probability distribution of a continuous random variable can be specified by giving a so-called *density function f*. Alternatively, you could also characterize the distribution by the corresponding *(cumulative) distribution function F*, which is defined as $F(t) := P(T \leq t)$. For example, a quite simple but typical choice in the survival analytic setting is the *exponential distribution*, whose density is given by

$$f(t) = \lambda \exp(-\lambda t), t \geq 0,$$

where λ is a fixed constant greater than 0.
Nevertheless, there are some additional quantities which are especially useful in time-to-event analysis.

Definition 2.1.

(i) $S(t) := P(T > t), t \in [0, \infty)$, is called *survival function*.

(ii) $\lambda(t) := \lim_{\Delta t \to 0} \frac{P(t \leq T < t + \Delta t | T \geq t)}{\Delta t}, t \in [0, \infty)$, is called *hazard function*.

(iii) $\Lambda(t) := \int_0^t \lambda(s)ds, t \geq 0$, is called *cumulative hazard function*.

These quantities have some basic properties which are stated in the following proposition.

Proposition 2.2 (Properties of $S(t), \lambda(t)$ and $\Lambda(t)$).

(i) $S(t) = 1 - F(t), t \geq 0$.

(ii) S is monotonically decreasing in t; $S(0) = 1$; $\lim_{t \to \infty} S(t) = 0$.

(iii) $\forall t \in [0, \infty) : \lambda(t) \geq 0$.

Proof. (i) follows directly from the definition of S. Applying the fundamental properties of a distribution function yields (ii) (note that $F(0) = 0$ because T is assumed to be nonnegative). (iii) follows from the fact that both the numerator and the denominator in the definition of λ are nonnegative. $\square$

Furthermore, S, λ and Λ are closely related to the density and the distribution function of T. Exactly speaking, the distribution of T can be specified by taking any of these quantities. In order to prove this, we first need a lemma that clarifies the relation between λ, f and S.

Lemma 2.3. *For all $t \in [0, \infty)$, we have*

$$\lambda(t) = \frac{f(t)}{S(t)}.$$

Proof. According to the definitions of λ and S and some basic probability theory, we have

$$\lambda(t) = \lim_{\Delta t \to 0} \frac{P(t \leq T < t + \Delta t | T \geq t)}{\Delta t} = \lim_{\Delta t \to 0} \frac{P(t \leq T < t + \Delta t)}{P(T \geq t)\Delta t}$$

$$= \lim_{\Delta t \to 0} \frac{F(t + \Delta t) - F(t)}{S(t)\Delta t} = S(t) \lim_{\Delta t \to 0} \frac{F(t + \Delta t) - F(t)}{\Delta t}.$$

Note that this limit is just the first derivative of F with respect to t. Since we assumed T to be a continuous random variable, this is equal to $f(t)$, and the proof is complete. $\qquad\square$

Example 2.4. To illustrate that lemma, let us turn back to our previous example. According to statement (i) in the proposition above, the survival function for the exponential distribution is given by

$$S(t) = 1 - F(t) = 1 - (1 - \exp(-\lambda t)) = \exp(-\lambda t).$$

The hazard function can be calculated using the previous lemma:

$$\lambda(t) = \frac{f(t)}{S(t)} = \frac{\lambda \exp(-\lambda t)}{\exp(-\lambda t)} = \lambda.$$

So, we see that the hazard for the exponential distribution is actually a constant.

Theorem 2.5.

(i) $\lambda(t) = -\frac{d \ln(S(t))}{dt}$.

(ii) $S(t) = \exp\left(- \int_0^t \lambda(x)dx\right)$.

Proof. As to the proof of statement (i), one just has to evaluate the expression on the right handside and apply some of the results from above:

$$-\frac{d \ln(S(t))}{dt} = -\frac{1}{S(t)}\frac{dS(t)}{dt} = \frac{1}{S(t)}\frac{dF(t) - 1}{dt} = \frac{f(t)}{S(t)} = \lambda(t).$$

(ii) follows directly by integrating and exponentiating the equation in (i). $\qquad\square$

So, to sum up our results once again, we now know that the probability distribution of a (continuous) random variable T - our survival time variable - can be specified by either the density f, the cumulative distribution function F, the survival function S or the hazard (cumulative hazard) function $\lambda(\Lambda)$. It often depends on the topic and the research question(s) which quantity you decide to take: Especially in survival analyses, you usually choose the survival or the hazard function, depending on which interpretation you are interested in: Roughly speaking, whereas the first one is a measure of the overall survival experience of the study population, the hazard function quantifies the risk of having the event at a certain time point. So, for example, if the investigator has a certain idea how the risk varies over time (e.g. increasing), the survival time distribution can be modelled by specifying a hazard function which reflects that behaviour.

Let us now turn to the problem of censoring, which is often a key feature of survival data. I have already illustrated this issue by giving some examples in the introduction. Now, I want to introduce some notation in order to be able to appropriately deal with censoring. Note that for the purposes of my master's thesis, I will restrict the focus on right censoring. For a discussion of other censoring schemes, I refer to Kalbfleisch and Prentice [14], pp. 78-83.

First, we should think about the question why we need to develop certain methods which allow for censoring. For instance, we could simply take the exact survival times and do some kind of linear or nonlinear regression with the survival time being the response and several predictors such as age, gender, etc. if your data comes from a medical context such as in our cancer study mentioned above. But to stay with this example, we would throw away a considerable amount of data - 110 out of 1,000. Apart from the fact that usually those follow-ups cost a lot of time and money, which then would have been spent in vain, one shouldn't forget about the information censored data provide: To take a rather extreme, but illustrative example, it's evident that there's a difference between 5 percent and 80 percent being still alive at the end of the study period. Of course, as the patients may be diagnosed at different time points throughout the study period (e.g. in 2005 and 2012), the censored observations as well as the information they provide can't be adequatly dealt with by just looking at that proportion. Thus, the main question is whether we can make use of their survival times somehow. Actually, the following chapters will show how this goal is achieved.

Until now, we have only considered a single random variable T representing survival time. However, when looking at real-life situations, we will have a sample of survival times, possibly with associated covariates and censoring. In order to formulate this setting mathematically, we introduce the following definitions:

Definition 2.6 (Censored sample (with covariates)). Let $(T_i^{ex})_{i=1}^n$ be a finite sequence of random variables with corresponding probability distributions $(P_i)_{i=1}^n$. Let $(C_i)_{i=1}^n$ be a finite sequence of random variables with densities $(g_i)_{i=1}^n$.
Let $T_i := \min\{T_i^{ex}, C_i\}$ and $\Delta_i := 1(T_i = T_i^{ex}), 1 \leq i \leq n$. Assume that $\{T_1^{ex}, T_2^{ex}, ..., T_n^{ex}, C_1, C_2, ..., C_n\}$ are independent.

(i) Then, the finite sequence $((T_i, \Delta_i))_{i=1}^n$ is called a *sample of right censored survival times* $(P_i)_{i=1}^n$. The elements of the sequence (C_i) are called *right censoring times*, the elements of (Δ_i) are called *censoring indicators*. T_i^{ex} is called *exact survival time*, T_i is called *observed survival time*, $1 \leq i \leq n$.

(ii) Let $(\mathbf{X_i})_{i=1}^n$ be a finite sequence of covariate vectors corresponding to the sequence of exact survival times. Assume that the probability distribution P_i depends on $\mathbf{X_i}, 1 \leq i \leq n$. The finite sequence $((T_i, \Delta_i, \mathbf{X_i}))_{i=1}^n$ is called a *sample of right censored survival times with covariate vectors* $(\mathbf{X_i})_{i=1}^n$.

Remark 2.7.

(i) The censoring scheme assumed above is usually referred to as "random censoring". In my master's thesis, I won't always state this fact explicitely because it is already included in my definition of a censored sample.

(ii) The term "observed survival time" does not mean that we have realizations here! By the word "observed", we emphasize the contrast to the exact survival times because the latter ones may not be observed for some subjects.

(iii) In order to stress the fact that the probability distributions P_i, specified by some quantity like the hazard function, depend on the covariate vectors $\mathbf{X_i}$, we sometimes write, for instance, $\lambda(t|\mathbf{X_i})$ instead of $\lambda_i(t)$.

(iv) Note that independence is required for the exact but not for the observed survival times! Nevertheless, the assumptions concerning independence in Definition 2.6 imply that the observed survival times $T_1, T_2, ..., T_n$ are independent, too. But be careful: Obviously, the T_is and the C_is are **not** independent!

Of course, these definitions look a bit technical, but there isn't a way to get around that effort since we need a proper notation for our further work.

To be honest, the definitions stated above can be made in a much more general way: There are other types of right censoring than the one I have defined. Exactly speaking, the above setting, that is, random (right) censoring, is a special case of a very general situation called *independent censoring*, which also covers censoring schemes where, for instance, the subjects are followed until the k-th event. But as I don't want to confuse the reader with too much terminology which is not illustrated by some examples, I refer to Kalbfleisch and Prentice [14], pp. 52-54, for the details concerning the generalizations of several censoring schemes. I just formulate what independent censoring means in order to make the reader see that the concept developed above indeed fulfils this independence assumption.

Definition 2.8 (Independent censoring). Let $((T_i, \Delta_i, \mathbf{X_i}))_{i=1}^n$ be a sample of survival times T_i with covariate vectors $(\mathbf{X_i})_{i=1}^n$, where Δ_i indicates if T_i is right-censored or not, i.e., $\Delta_i = 1$ if subject i experiences the event of interest within the study period and $\Delta_i = 0$ otherwise, $1 \leq i \leq n$. The underlying right-censoring mechanism is called *independent* if the values of the hazard function that apply to individuals on trial at each time $t > 0$ are the same as those that would have been applied if there had not been censoring. In other words, for each $t > 0$ and $i \in \{1, 2, ..., n\}$, the following equation must be satisfied:

$$\lim_{\Delta t \to 0} \frac{P(t \leq T_i^{ex} < t + \Delta t | \mathbf{X_i} = \mathbf{x_i}, T_i^{ex} \geq t)}{\Delta t}$$
$$= \lim_{\Delta t \to 0} \frac{P(t \leq T_i^{ex} < t + \Delta t | \mathbf{X_i} = \mathbf{x_i}, T_i^{ex} \geq t, Y_i(t) = 1)}{\Delta t}, \quad (2.1)$$

where $Y_i(t) = 1$ if subject i is at risk at time t and $Y_i(t) = 0$ otherwise.

Proposition 2.9. *Let $((T_i, \Delta_i, \mathbf{X_i}))_{i=1}^n$ be a sample of right-censored survival times with covariate vectors $(\mathbf{X_i})_{i=1}^n$ as in Definition 2.6. Then, the underlying censoring scheme is independent.*

Proof. To begin with, in the setting of Definition 2.6, $Y_i(t) = 1$ is equivalent to $T_i^{ex} \geq t \wedge C_i \geq t$. Thus, we can write

$$P(t \leq T_i^{ex} < t + \Delta t | \mathbf{X_i} = \mathbf{x_i}, T_i^{ex} \geq t, Y_i(t) = 1)$$
$$= P(t \leq T_i^{ex} < t + \Delta t | \mathbf{X_i} = \mathbf{x_i}, T_i^{ex} \geq t, C_i \geq t).$$

But we have also assumed that T_i and C_i are independent. So we can drop the condition on C_i and get

$$P(t \leq T_i^{ex} < t + \Delta t | \mathbf{X_i} = \mathbf{x_i}, T_i^{ex} \geq t, C_i \geq t)$$
$$= P(t \leq T_i^{ex} < t + \Delta t | \mathbf{X_i} = \mathbf{x_i}, T_i^{ex} \geq t).$$

All in all, we have shown that

$$\lim_{\Delta t \to 0} \frac{P(t \leq T_i^{ex} < t + \Delta t | \mathbf{X_i} = \mathbf{x_i}, T_i^{ex} \geq t, Y_i(t) = 1)}{\Delta t}$$

is equal to

$$\lim_{\Delta t \to 0} \frac{P(t \le T_i^{ex} < t + \Delta t | \mathbf{X_i} = \mathbf{x_i}, T_i^{ex} \ge t)}{\Delta t},$$

which means that the censoring scheme is independent.   □

Again, we emphasize that in the definition above, some assumptions of random right censoring are relaxed. However, the random censorship setting is sufficiently general for many practical applications, and we have seen that it fulfils the independence assumption, so considering this censoring scheme only will do for the remaining part of my master's thesis. But, one has to keep in mind that even the assumptions required for independent censoring can be violated in situations which are quite often encountered, as demonstrated in the remark below.

Remark 2.10. To further illustrate the concept of independent censoring, it may be instructive to look at an example where the censoring scheme is **not** independent. Just think of a medical study on the time to death of seriously ill patients, where some surgical intervention is taken as starting point. Let us assume that for practical reasons, patients are discharged from hospital when a certain score, which measures the degree of recovery, exceeds a predefined value. After discharge from hospital, the patients will not be followed any more, so, to cut it short, they will be right censored if their score is fairly high. However, this censoring scheme is not independent: If you take a look at the right handside of equation (2.1), you see that the conditions $\mathbf{X_i} = \mathbf{x_i}$ (which involves the quite large score value mentioned above) and $Y_i(t) = 1$ (which in particular means that individual i is not censored at time t) cannot hold simultaneously since we said that an individual with a high score will be censored. Consequently, the right handside of equation (2.1) is not well defined. Thus, the censoring scheme is not independent.

Example 2.11. Now, the terminology introduced above is illustrated by looking at our epilepsy dataset. Remember that we decided to take only a subset of the data, namely the so-called incidence cohort, which consist of patients who came to the clinic within one year after their first seizure. The date of their first visit at the clinic, which is referred to as *behandlungsbeginn*, is considered to be the time origin. The corresponding censoring time, call it C_i, is the span between *behandlungsbeginn* and Dec 31, 2010. Recall that we don't have to care about patients lost to follow up, as stated previously.

In addition to that, although the subjects are scheduled for subsequent examinations, e.g. to find out if their health status has changed, all the covariates (gender etc.) we are interested in are measured at the time of entry and remain fixed during the entire study period. Summing up, all we need to know for our analysis is measured at the time of entry into the study (the first visit at the clinic), and thus, given the vector of covariates $\mathbf{x_i}$ and the value of the variable *behandlungsbeginn*, the associated censoring time C_i is fixed, as it is given by the difference between end of the study and date of first visit. As we deal with censoring times being constants, we obviously have a special case of (random) right censoring here. Particularly, our underlying censoring scheme for the epilepsy dataset is independent.

By the way, it should be mentioned that we could also choose a slightly different setting. Instead of looking only at the incidence cohort, we could take the whole dataset and define the value of the variable *anfallsbeginn*, which represents the date of a patient's first epileptic seizure, as starting point. Of course, we must then account for the fact that for some subjects, the time span between *anfallsbeginn* and *behandlungsbeginn* is quite large, which may be a source of bias since those patiens had already survived for a fairly long time until they were enrolled. Actually, there are survival analytic methods available that account for the phenomenon of so-called *delayed entry* or *left truncation*. Note that this is different from *left censoring*: When left censoring occurs, we have at least partial information on the survival times insofar as we know that they do not exceed a certain value (just remember the answer "I drank alcohol, but I cannot remember the exact age" in our hypothetical study presented in the introductory chapter). However, left truncation does not give us further information about the subjects which experienced the event of interest prior to the entry time. For example, imagine a study being conducted at a retirement center, with the age of the subjects as time to event (Klein and Moeschberger [16], pp. 16-17). Obviously, people have to survive to a sufficient age to be allowed to enter the retirement center. Thus, we do not know anything about the survival experience of people who died prior to that certain age. Furthermore, you can also see what's the main problem with the analysis of left truncated data: Although there are methods available which enable the investigator to properly deal with that type of data, we eventually get conditional quantities. But as you can see when looking at the epilepsy data, the time spans between onset of seizures and the first visit at the clinic range from a few days up to ten or twenty years. Thus, we would only be able to get statements like "if a patient with certain characteristics survives 10 years, his or her survival probability will be...". However, we want to find out how life expectancies change over time, from the date of first visit at the clinic up to 20 years after diagnosis, as will be discussed in chapter 5. Especially for the early years, truncation would make such statements impossible, and therefore, I think it is a better decision to stay with the incidence cohort approach.

In addition to independence, there is another property of censoring schemes which must not be confused with the first one:

Definition 2.12 (Noninformative censoring). Let $((T_i, \Delta_i))_{i=1}^n$ be a sample of censored survival times as in Definition 2.6. Let $(f_i(\theta)), (g_i)$ be the densities of (T_i^{ex}) and (C_i), respectively, where θ is a vector of parameters. The underlying censoring scheme is said to be *noninformative* if the densities g_i do not involve θ.

Remark 2.13. A completely analogous definition can be stated if we additionally incorporate covariate vectors $(\mathbf{X_i})$ in our model. We will see in the next chapter that especially in a regression context, the noninformativity property is of importance when constructing the likelihood function.

Now, after this rather extensive account of censoring, we turn back to the key survival analytic quantities we have defined at the beginning of this chapter. So far, we haven't talked about the estimation of the survival and the (cumulative) hazard function based on a sample of censored survival times. First, we introduce a very basic, but important approach. Let us consider a finite sequence $((t_i, \delta_i))_{i=1}^n$ of observations arising from a homogeneous population with survival function S_0. Let us apply some kind of re-indexing to our data such that the exact survival times are denoted by $t_1, t_2, ..., t_k$ and sorted in ascending order. Furthermore, we assign double-indices to the remaining $n - k$ censored observations, thus indicating which time interval they belong to. More to the point, we denote the m_j censoring times in the interval $[t_j, t_{j+1})$ by $t_{j1}, t_{j2}, ..., t_{jm_j}, j \in \{0, 1, ..., k\}$, where we set $t_0 := 0$ and $t_{k+1} := \infty$ (although we don't need these quantities right now, this notation is very useful for the derivations carried out later in this chapter). Let d_j denote the number of events at t_j and $n_j := d_j + m_j + d_{j+1} + m_{j+1} + ... + d_k + m_k$ the number of subjects at risk at a time just prior to t_j, respectively, $0 \leq j \leq k$. Now, we want to estimate the unknown survival function S_0 (or, equivalently, any other survival analytic quantity which uniquely determines the underlying distribution). There are two very popular approaches which are introduced in the following definitions.

Definition 2.14 (Kaplan-Meier estimate (Kaplan and Meier [15])). The function $\hat{S} : \mathbb{R}^+ \to [0, 1]$ defined by

$$\hat{S}(t) := \prod_{\{j : t_j \leq t\}} \left(1 - \frac{d_j}{n_j}\right)$$

is called *Kaplan-Meier estimate* or *Product-Limit estimate* of the survival function S_0.

Definition 2.15 (Nelson-Aalen estimate (Nelson [18], Aalen [1])). The function $\hat{\Lambda} : \mathbb{R}^+ \to \mathbb{R}^+$ defined by

$$\hat{\Lambda}(t) := \sum_{\{j : t_j \leq t\}} \frac{d_j}{n_j}$$

is called *Nelson-Aalen estimate* of the cumulative hazard function Λ_0.

Remark 2.16.

(i) Note that the Kaplan-Meier estimate does not necessarily reduce to 0: For example, if

$$i_0 := \underset{i}{argmax}\, t_i, \delta_{i_0} = 0,$$

that is, the largest survival time is a censored observation, we have $d_i/n_i \neq 1$ for all $i = 1, 2, ..., k$. In this case, we take $\hat{S}(t)$ to be undefined for $t > t_{i_0}$. Alternatively, we could, for instance, set $\hat{S}(t) = 0$ for $t > t_{i_0}$. But, when doing so, one always has to keep in mind that we thus introduce some additional assumptions concerning the underlying survival process. As to the alternative solution given above, setting the estimate to 0 means that we assume nobody can survive beyond t_{i_0}.

(ii) Both quantities defined above involve $d_i/n_i, 1 \leq i \leq n$. These quotients can be regarded as very natural hazard estimates since they represent the observed deaths relative to the number of subjects at risk.

(iii) As to the general outlook of the functions $\hat{S}$ and $\hat{\Lambda}$, it is obvious from the formulas that they both are step functions with jumps at the observed exact survival times $t_1, t_2, ..., t_k$. One should keep in mind that the size of these jumps is not only determined by the number of deaths d_j, but also by the number of censored observations in the interval $[t_j, t_{j+1})$ because the latter quantities contribute to n_j. Thus, we now see how censoring is accounted for in the estimates defined above.

(iv) Due to the results stated in Theorem 2.5, we can as well use $-ln(\hat{S})$ as an estimate of Λ_0 and $exp(-\hat{\Lambda})$ to estimate S_0, respectively.

For now, we restrict our focus to the Kaplan-Meier estimate. We have already stated in the previous remarks that the components of $\hat{S}(t)$

are basically made up by some very natural hazard estimates. Therefore, the Kaplan-Meier estimate may be a somehow reasonable and "good" choice. However, we should further clarify the properties of $\hat{S}(t)$.

At first, we take a look at a special case that arises if there aren't any censored observations in the sample.

Proposition 2.17. *Let $((t_j, \delta_j))_{j=1}^n$ be a sequence of observations with $\delta_j = 1$ for all $j \in \{1, 2, ..., n\}$. Furthermore, we assume that $t_1 < t_2 < ... < t_n$. Then, we have*

$$\hat{S}(t) = 1 - \hat{F}(t)$$

for all $t \geq 0$, where $\hat{F}$ denotes the empirical CDF.

Proof. Due to the fact that $\delta_j = 1$ for all $j \in \{1, 2, ..., n\}$, we have $n_j - d_j = n_{j+1}$. This yields

$$\hat{S}(t) = \prod_{\{j:t_j \leq t\}} \left(1 - \frac{d_j}{n_j}\right) = \prod_{\{j:t_j \leq t\}} \frac{n_j - d_j}{n_j} = \prod_{\{j:t_j \leq t\}} \frac{n_{j+1}}{n_j} = \frac{n_{j_0+1}}{n_0},$$

where $n_{j_0+1} := max\{j \in \{1, 2, ..., n\} : t_j \leq t\}$. Now, note that $n_0 = n$ and $n - n_{j_0+1}$ is equal to the number of events in the interval $[t_0, t_{j_0+1})$, which is in turn $\#\{j : t_j \leq t\}$. The second statement again follows from the assumption of no censoring. All in all, we thus get

$$1 - \hat{S}(t) = 1 - \frac{n_{j_0+1}}{n_0} = \frac{\#\{j : t_j \leq t\}}{n},$$

which completes the proof. $\qquad\square$

Next, we recall an important result concerning the empirical CDF.

Proposition 2.18. *Let $\hat{F}$ be the empirical CDF based on a sample $x_1, x_2, ..., x_n$ arising from a distribution with (unknown) CDF F_0. Let*

$$L(F) := \prod_{i=1}^n (F^+(x_i) - F^-(x_i))$$

denote the nonparametric likelihood for a given CDF F. Then, if we arbitrarily choose a CDF F, the following inequality holds:

$$L(F) < L(\hat{F})$$

In other words, $\hat{F}$ is the (unique) nonparametric maximum likelihood estimate of F_0.

Proof. See Owen [20], p. 8. □

Since the MLE property is invariant under monotonic transformations, we can combine the two propositions stated above in order to get the result that in the special case where there is no censoring, the Kaplan-Meier estimate $\hat{S}$ is the nonparametric MLE of the true survival function S_0. But, if we allow for censored observations, the question arises whether the MLE property is still valid or not. The following theorem tells us that indeed, $\hat{S}$ is a nonparametric MLE of S_0 in this general situation, too. With the notations introduced right above Definition 2.14 and the assumption that the underlying censoring scheme is independent, the likelihood function for an arbitrarily chosen survival function S is given as

$$L(S) = \prod_{j=0}^{k} \left(\left(S(t_j^-) - S(t_j^+) \right)^{d_j} \prod_{l=1}^{m_j} S(t_{jl}) \right). \tag{2.2}$$

The first factor can be explained by the very same arguments as for the "usual" nonparametric likelihood defined above. The latter product results from the assumption of independent censoring: For a censored observation, all we know is that the actual, unobserved survival time exceeds the censored one. This statement can be proven by using counting process techniques, see Kalbfleisch and Prentice [14], pp. 193-196.

Now, we are ready to state the theorem mentioned above (for the following proof, see Kalbfleisch and Prentice [14], pp. 15-16):

Theorem 2.19 (MLE property of the Kaplan-meier estimate). *With the assumptions and notational conventions from above, the Kaplan-Meier estimate $\hat{S}$ is a nonparametric maximum likelihood estimate of the (unknown) survival function S_0, i.e., $L(S) \leq L(\hat{S})$ for any survival function S.*

Proof. To start with, we derive two conditions the nonparametric MLE S_0 necessarily has to fulfil. Firstly, if $S(t_j^-) = S(t_j^+)$ holds for a $j \in \{1, 2, ..., k\}$, the likelihood given in (2.2) vanishes. Consequently, such a S would certainly not maximize the likelihood. So, a nonparametric MLE, call it $\tilde{S}$, must be discontinuous at the exact survival times $t_1, t_2, ..., t_k$. Secondly, recall that a survival function is always monotonically decreasing in t. Therefore, if we now take a look at maximizing the values of $\tilde{S}$ at the censored observations (i.e., the latter part of (2.2)), we see that we have to set $\tilde{S}(t_{jl}) = \tilde{S}(t_j)$ because we have chosen the double-indices such that $t_j \leq t_{jl}$, $j \in \{1, 2, ..., k\}, l \in \{1, 2, ..., m_j\}$. So, summing up, we want to find a discrete survival function $\tilde{S}$ with hazard components $\tilde{\lambda}_1, \tilde{\lambda}_2, ..., \tilde{\lambda}_k$ at the exact survival times $t_1, t_2, ..., t_k$. In particular, we therefore get

$$\tilde{S}(t) = \prod_{\{l:t_l \leq t\}} (1 - \tilde{\lambda}_l),$$

where $t \geq 0$ (Kalbfleisch and Prentice [14], p. 9). Now, we plug in this formula in (2.2) and obtain

$$L(\tilde{S}) = \prod_{j=0}^{k} \left(\left(\tilde{S}(t_j^-) - \tilde{S}(t_j^+) \right)^{d_j} \prod_{l=1}^{m_j} \tilde{S}(t_{jl}) \right)$$

$$= \prod_{j=1}^{k} \left(\left(\prod_{l=1}^{j-1}(1 - \tilde{\lambda}_l) - \prod_{l=1}^{j}(1 - \tilde{\lambda}_l) \right)^{d_j} \left(\prod_{l=1}^{j}(1 - \tilde{\lambda}_l) \right)^{m_j} \right)$$

$$= \prod_{j=1}^{k} \left(\left(\prod_{l=1}^{j-1}(1 - \tilde{\lambda}_l)(1 - 1 + \tilde{\lambda}_j) \right)^{d_j} \prod_{l=1}^{j}(1 - \tilde{\lambda}_l)^{m_j} \right)$$

$$= \prod_{j=1}^{k} \tilde{\lambda}_j^{d_j} \prod_{l=1}^{j-1} (1 - \tilde{\lambda}_l)^{d_j} \prod_{l=1}^{j} (1 - \tilde{\lambda}_l)^{m_j}$$

$$= \prod_{j=1}^{k} \tilde{\lambda}_j^{d_j} (1 - \tilde{\lambda}_j)^{n_j - d_j} \tag{2.3}$$

So, we see that maximizing (2.2) on the space of all survival functions eventually reduces to the problem of finding $\tilde{\lambda}_1, \tilde{\lambda}_2, ..., \tilde{\lambda}_k$ which maximize (2.3). When looking at the latter formula, we see that the factors are proportional to the PDFs of binomial distributions with parameters $\tilde{\lambda}_j, j = 1, 2, ..., k$. As we don't care about multiplicative constants in ML estimation, we can use what we know about the binomial ML estimate and maximize each of the factors by setting $\tilde{\lambda}_j = d_j/n_j, j = 1, 2, ..., k$. Due to the fact that maximizing the product in (2.3) is equivalent to maximizing each of the factors, we thus have a nonparametric MLE $\tilde{S}$ of S_0, namely

$$\tilde{S}(t) := \prod_{\{j : t_j \leq t\}} \left(1 - \frac{d_j}{n_j} \right),$$

which agrees exactly with the definition of the Kaplan-Meier estimate $\hat{S}$. $\qquad\square$

As we won't go into further details concerning the properties of the Kaplan-Meier estimator, we just state some results in a rather informal way and continue with an illustration of the theory outlined above by calculating $\hat{S}$ for the epilepsy dataset.

Remark 2.20. Under relatively mild conditions, the following properties of the Kaplan-Meier estimator can be shown:

(i) $\hat{S}$ is uniformly consistent for S_0 on the time interval $[0, \tau]$, where τ denotes the end of the study period.

(ii) For fixed t, the distribution of $\hat{S}(t)$ can be approximated by a normal distribution with mean $S_0(t)$ and estimated variance

$$\widehat{Var}(\hat{S}(t)) = \hat{S}^2(t) \sum_{\{j:t_j \leq t\}} \frac{d_j}{n_j(n_j - d_j)}.$$

This expression for the asymptotic variance is known as *Greenwood's formula*.

For detailed information concerning these results and the corresponding proofs, we refer to Kalbfleisch and Prentice [14], pp. 17-18, 167-171.

Example 2.21. In Figure 2.1, we see the Kaplan-Meier estimates for female and male patients, respectively. To create such a plot, we basically used the **survreg** function contained in the **survival** package (Therneau [26]). Throughout the entire time range, the female patients have a slightly better chance of surviving than the males. By the way, there are formal hypothesis tests for the comparison of survival curves available, but we shall not discuss this topic here as it is covered by every standard textbook such as Klein and Moeschberger [16], pp. 205-214 or Kalbfleisch and Prentice [14], pp. 20-23.

What we also see in the plot is the fact that, as mentioned above, the Kaplan-Meier estimate is undefined beyond the largest observed time if this is a censored one. Indeed, the largest observed survival times for female and male are 14245 d (approx. 39 y) and 14297 d (about 39.14 y), respectively, which are both censored observations. As a consequence, we can see in the plot that the step functions representing the KM estimates are not drawn beyond these time points.

However, from a medical point of view, it would be hardly reasonable if we only accounted for the gender of the patients. For example, we have seen in the introductory chapter that the range of the variable *age at diagnosis* is quite large. Therefore, taking further characteristics of the patients into consideration will certainly give a more realistic

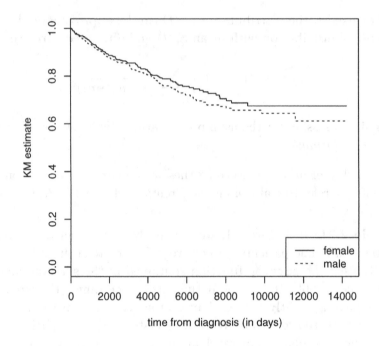

Figure 2.1: Kaplan meier estimates for female and male patients

and reliable picture of the study cohort's survival experience. So, let's calculate age- and gender-specific KM estimates. But, due to the large amount of possible values of *age at diagnosis*, it would make no sense to consider each possible combination of *gender* and *age at diagnosis*. Thus, we must divide the observations into distinct age groups instead. E.g., building up the four age groups <25, 25-45, 46-65 and >65 seems to be a plausible choice since the lenghts of these intervals are approximately equal. Moreover, we thus get 8 possible combinations of the two covariates we want to take into consideration, which is a number that should still work quite well when visualising

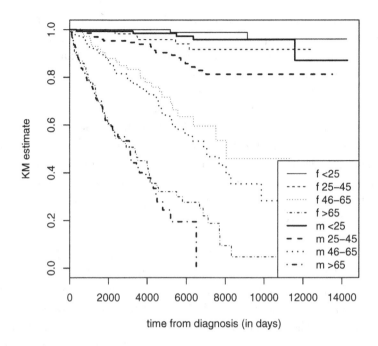

Figure 2.2: Kaplan meier estimates stratified according to age and gender

the corresponding KM estimate graphs in one single plot. The results are displayed in Figure 2.2.

To begin with, it is small surprise that the survival probability estimates decrease when we proceed from the first to the last age group. What's more interesting is the fact that the estimated survival curve for $m > 65$ eventually arrives at 0 since in contrast to the other groups, the largest observed survival time is not censored. In this case, the Kaplan-Meier estimate is well defined on the whole positive x-axis.

Furthermore, as to the comparisons between male and female, the graphs for the age groups 46 − 65 and > 65 look quite nice and are thus easily interpretable: The women are better off than the men. This seems to hold true for the patients between 25 and 45, too. However, whereas the graph for the men of this age group provides sufficient information (we have 22 exact survival times, meaning that the KM estimate is a step function with 22 jumps) about the general outlook of the survival function, the corresponding estimate for the female patients is based on only 6 exact observations. Likewise, we can hardly make any meaningful statements based on the two graphs for the first age group since the number of jumps is 2 for female and 5 for male, respectively.

So, we see a very important point here: The fact that the KM estimate - by the way, the same problem would arise if we used the Nelson-Aalen approach - is discontinuous only at the observed exact survival times can cause serious problems, especially when we want to take several characteristics of the patients into account. Of course, we could choose the age intervals in a way such that we have a sufficiently large number of exact observations in each group. But, for example, if we combine the first and the second age groups to one single group for male and female patients, respectively, it is questionable if the results are still useful as far as their interpretability is concerned: Most likely, there is a striking difference between the survival experiences of a newborn and a 45-year-old patient. But, when merging these patients into one single group, we implicitly assume that such a difference does not exist. Moreover, recall that the patients were diagnosed in a certain year between 1970 and 2010. When looking at the corresponding population life tables for Tyrol (see chapter 5), one observes a great change in the survival experience in the course of those years. So, apparently, besides the covariates *gender* and *age at diagnosis*, we have to account for *year of diagnosis*, too. So, we must split up our original data (and, thus, especially the uncensored observations) into even more subsets to get reliable results. However, this certainly leads to further complications.

3 Regression models for survival data

In this chapter, we, at first, give a short explanation why regression models are needed for many different types of survival data. Then, we introduce two very popular classes of survival analytic regression models and discuss the relationship between them. When examining the epilepsy dataset, we repeatedly use a particular parametric model called the *Weibull model*, which is therefore defined in this chapter. Finally, we focus on the topic of parameter estimation. As in the previous chapter, we will mostly follow the textbooks of Klein and Moeschberger [16], pp. 74-78, 243-245, 253-254, 395-400 as well as Kalbfleisch and Prentice [14], pp. 31-34, 40-45, 52-57, 68-70.

As mentioned before, we often encounter situations when it makes hardly any sense to claim homogeneity of the study cohort in terms of survival time. For example, it is reasonable to assume that the risk of death varies according to differences in gender and age. Therefore, we may split the original data into several subgroups and investigate each of them separately, as we did in the last example of the previous chapter. By the way, we could also choose a parametric approach instead of using the nonparametric Kaplan-Meier estimator: There is a wide variety of parametric models available which are especially suited for analyzing survival data. For example, we could take the exponential distribution which was already touched in the introductory chapter. Of course, the parameter estimation is made more complicated by the fact that we have to account for censoring, but nevertheless, it has been shown that maximum likelihood methods can be extended to this situation. So, the only thing left to check concerns the question if the model we have chosen fits the data well. The answer can be given by using procedures which are similar to approaches that are well known from other statistical branches.

However, I want to discuss a slightly different and more complicated method now. As you will see, the approach outlined above is, in a way, a special case thereof, which can be handled a bit easier. To start with, let survival time be denoted by a continuous non-negative random variable T. Now, our goal is to model the relationship between T (or an associated survival analytic quantity) and a vector of covariates $\mathbf{X} := (X_1, X_2, ..., X_p)'$. We assume that these covariates are *time-independent*, which means that their value doesn't depend on time t.[1] Generally speaking, there are two main classes of regression models which can be used in this setting:

Definition 3.1 (Relative risk and accelerated failure time models). Let $\boldsymbol{\alpha} := (\alpha_1, \alpha_2, ..., \alpha_p)', \boldsymbol{\beta} := (\beta_1, \beta_2, ..., \beta_p)'$ be vectors of regression coefficients.

(i) A *relative risk model* or *Cox model* (Cox [6]) is a regression model specified by

$$\lambda(t|\mathbf{X}) = \lambda_0(t) \exp(\mathbf{X}'\boldsymbol{\beta}), \qquad (3.1)$$

where λ_0 denotes an arbitrary hazard function for a non-negative continuous random variable, the so-called *baseline hazard*.

(ii) Let $Y := ln(T)$. Let us assume the linear model

$$Y = \mu + \mathbf{X}'\boldsymbol{\alpha} + W, \qquad (3.2)$$

where W denotes an arbitrary continuous random variable, the *error variable*. This model is called an *accelerated failure time model*.

---

[1] It is important to note that even if the covariates are *time-dependent*, there are methods available which enable the researcher to properly deal with this issue: For example, the relative risk model can be suitably modified to accomodate time-dependent covariates, see Kalbfleisch and Prentice [14], pp. 96-99.

Remark 3.2.

- As to the relative risk model, it is important to keep in mind that the covariates incorporated in the model need not be time-independent, as already mentioned in a footnote above. However, if they are, this class of models is often named *proportional hazards model*. The reason is as follows: Suppose that we have two individuals with time-independent covariate vectors $\mathbf{x_1}$ and $\mathbf{x_2}$. In other words, the covariate values are measured before or at time 0 and remain constant over the entire time range. Then, for any $t \geq 0$, the following line holds:

$$\frac{\lambda(t|\mathbf{x_1})}{\lambda(t|\mathbf{x_2})} = \frac{\lambda_0(t)\exp(\mathbf{x_1}'\boldsymbol{\beta})}{\lambda_0(t)\exp(\mathbf{x_2}'\boldsymbol{\beta})} = \frac{\exp(\mathbf{x_1}'\boldsymbol{\beta})}{\exp(\mathbf{x_2}'\boldsymbol{\beta})}.$$

So, we arrive at the fact that the hazards of two arbitrarily chosen individuals are proportional at any time point t since the expression on the right handside of the equation above is independent of t.

- To see why the models of class (ii) are called accelerated failure time models, let us apply a back-transformation (for the following proof, see Klein and Moeschberger [16], pp. 46-49, and Kalbfleisch and Prentice [14], p. 44): When exponentiating (3.2), we get

$$T = \exp(\mu)\exp(\mathbf{X}'\boldsymbol{\alpha})U,$$

where $U := exp(W)$ is a non-negative continuous random variable. Let S_U denote the corresponding survival function. Now, we show that the survival function S corresponding to T can be expressed in terms of S_U: For any $t \geq 0$, we have

$$\begin{aligned} S(t|\mathbf{X}) &= P(T > t|\mathbf{X}) = P(\exp(\mu)\exp(\mathbf{X}'\boldsymbol{\alpha})U > t|\mathbf{X}) \\ &= P(U > t\exp(-\mu)\exp(-\mathbf{X}'\boldsymbol{\alpha}), \mathbf{X}) \\ &= S_U(t\exp(-\mu)\exp(-\mathbf{X}'\boldsymbol{\alpha})). \end{aligned} \qquad (3.3)$$

Due to the fact that $S_U(\exp(-\mu)t) = S(t|\mathbf{X} = \mathbf{0})$, we can interpret S_U as a baseline survival function, say, S_0.[2] Thus, the final result of the calculations above tells us that the covariates either "accelerate" or "decelerate" the survival time t. This is a striking difference to the relative risk model, which can be best seen when we apply the ln to equation (3.3) and differentiate with respect to t to get

$$\lambda(t|\mathbf{X}) = -\frac{d\ln(S(t|\mathbf{X}))}{dt} = -\frac{d\ln(S_0(t\exp(-\mu)\exp(-\mathbf{X}'\alpha)))}{dt}$$

$$= -\frac{-f_0(t\exp(-\mu)\exp(-\mathbf{X}'\alpha))(\exp(-\mu)\exp(-\mathbf{X}'\alpha))}{S_0(t\exp(-\mu)\exp(-\mathbf{X}'\alpha))}$$

$$= \exp(-\mu)\exp(-\mathbf{X}'\alpha)\lambda_0(t\exp(-\mu)\exp(-\mathbf{X}'\alpha)).$$

So, we see that in contrast to (3.1), where the covariate term acts multiplicatively **on the hazard**, we have an additional multiplicative acceleration/deceleration effect **on time** in the AFT model (3.2).

- It must be stated that in the past decades, many other types of regression models have been proposed. Apart from the rather obvious fact that we can replace the exponential link function in (i) by almost every other function, it is also possible to consider, for example, so-called *additive hazard rate models*. As the name suggests, the hazard function is modeled by the sum of a baseline hazard and the standard inner product of the covariate and the regression coefficient vector. Despite the assumption that the hazard can be decomposed in this way, this class of models is quite flexible: For instance, the regression coefficients as well as the covariates are allowed to be time-varying. Details concerning this approach can be found in Klein and Moeschberger [16], pp. 329-333, 346-348.

- Moreover, assuming a certain structure of the hazard function can make sense in many situations because often, one has quite a good

---

[2] The scaling constant $\exp(-\mu)$ does not matter here. By the way, don't confuse this baseline survival function with the true survival function S_0 in the previous chapter.

idea of what the hazard looks like and how it is influenced by taking several covariates into account. However, especially if the hazard rates are not of primary interest, it may be better to use other survival analytic quantities. For example, one could set up a model analogously to (i), but replace the (baseline) hazard function by the (baseline) mean residual life, which is defined in chapter 5. More information about this fairly recent approach can be found in Sun and Zhang [25].

Specific parametric models belonging to the two classes introduced above can be quite easily set up by choosing certain functional forms for the baseline hazard and the distribution of W, respectively. This is demonstrated in the following example:

Example 3.3. We consider a special relative risk model which is defined by setting

$$\lambda_0(t) = \lambda p t^{p-1},$$

where λ and p are positive constants. This model is called *Weibull regression model*, because it was proposed by W. Weibull in 1951 (see Weibull [29]). Interestingly, in this paper, the main focus lies on the discussion of the wide range of applications in the industrial branch (e.g., examination of the strength of steel), whereas questions concerning human life are only peripherically touched.

By the way, we could also define the hazard function $\lambda(t)$ itself as $\lambda(t) := \lambda p t^{p-1}$. This hazard function is then said to specify a model called *Weibull model*. Note that in contrast to the definition above, we don't have a regression part here! This difference is stressed by the additional word "regression" in the name of the first model. Nevertheless, in the latter model, we can also account for covariates by splitting the data into subgroups and fit separate Weibull models for these subsets, as mentioned in the introductory part of this chapter.

Once we know the hazard, the other survival analytic key quantities can be derived using the results from chapter 2. For instance, the survival function for a Weibull regression model is then given as

$$S(t) = exp\left(-\int_0^t \lambda(s)ds\right)$$

$$= exp\left(-\int_0^t \lambda p s^{p-1} \exp(\mathbf{x}'\boldsymbol{\beta})ds\right)$$

$$= exp\left(-\lambda t^p exp(\boldsymbol{\beta}'\mathbf{x})\right). \tag{3.4}$$

Interestingly, we can specify the Weibull regression model via the accelerated failure time approach, too. To see this, consider a slightly different notation of a linear model for $Y := ln(T)$, namely

$$Y = \mu + \mathbf{X}'\boldsymbol{\alpha} + \sigma W,$$

where the only difference compared to (3.2) is the occurrence of an additional scaling factor σ. Now, we set the density f_W of the error variable W to

$$f(w) := \exp(w - \exp(w)), w \geq 0,$$

which is known as the *extreme value distribution*.
By integration, we get the corresponding survival function as

$$S_W(w) = \int_w^\infty f_W(v)dv = \int_w^\infty exp(v - exp(v))dv = exp(-exp(w)).$$

Because $P(Y \lessgtr y) = P(W \lessgtr \sigma^{-1}(y - \mu - \mathbf{x}'\boldsymbol{\alpha}))$, the density and survival function for Y are given by

$$f_Y(y) = \frac{1}{\sigma} exp\left(\frac{y - \mathbf{x}'\boldsymbol{\alpha} - \mu}{\sigma} - exp\left(\frac{y - \mathbf{x}'\boldsymbol{\alpha} - \mu}{\sigma}\right)\right),$$

$$S_Y(y) = exp\left(-exp\left(\frac{y - \mathbf{x}'\boldsymbol{\alpha} - \mu}{\sigma}\right)\right).$$

Note that the constant $1/\sigma$ appears in the density, but not in the survival function because we must differentiate with respect to y to get f_Y. Now, simplifying $S_Y(y)$ yields

$$S_Y(y) = exp\left(-exp\left(\frac{y - \mathbf{x}'\boldsymbol{\alpha} - \mu}{\sigma}\right)\right)$$

$$= exp\left(-exp\left(\frac{y}{\sigma}\right)\exp\left(\frac{-\mathbf{x}'\boldsymbol{\alpha} - \mu}{\sigma}\right)\right) = exp\left(-exp\left(\frac{y}{\sigma}\right)C\right),$$

where

$$C := \exp\left(\frac{-\mathbf{x}'\boldsymbol{\alpha} - \mu}{\sigma}\right)$$

does not involve y. Thus, by applying the strictly monotonic transformation $T = exp(Y)$, we get

$$S_T(t) = P(Y \geq ln(t))$$

$$= exp\left(-exp\left(\frac{ln(t)}{\sigma}\right)C\right)$$

$$= exp\left(-exp\left(ln\left(t^{1/\sigma}\right)\right)C\right)$$

$$= exp\left(-t^{1/\sigma}C\right)$$

$$= exp\left(-t^{1/\sigma}\exp\left(\frac{-\mathbf{x}'\boldsymbol{\alpha} - \mu}{\sigma}\right)\right).$$

Now, if we set

$$\lambda := exp\left(-\frac{\mu}{\sigma}\right), p := \frac{1}{\sigma}, \beta_i := -\frac{\alpha_i}{\sigma},$$

we arrive at the desired result, namely

$$S_T(t) = exp\left(-\lambda t^p exp(\boldsymbol{\beta}'\mathbf{x})\right),$$

which is equal to the survival function derived by the relative risk approach stated in (3.4).

In the previous example, we have seen an instance of a model which has a relative risk as well as a accelerated failure time representation. Naturally, one may ask if we could find other models which have this property, too. Suprisingly, this is not possible, as stated in the following proposition.

Proposition 3.4. *Let λ_R and λ_A be hazard functions arising from a relative risk model and an accelerated failure time model, respectively. If we assume that $\lambda_R(t) = \lambda_A(t)$ for all $t \geq 0$, i.e., the model for survival time T has a relative risk as well as an accelerated failure time representation, we have that for any $t \geq 0$, the following equation holds:*

$$\lambda_{R0}(t) = \lambda p t^{p-1},$$

where λ_{R0} denotes the baseline hazard of the relative risk model and λ, p are positive real constants.

Proof. According to the assumption $\lambda_R(t) = \lambda_A(t)$ and Remark 3.2, we have

$$\lambda_{R0}(t) \exp(\mathbf{X}'\boldsymbol{\beta}) = \lambda_{A0}(t \exp(-\mu) \exp(-\mathbf{X}'\boldsymbol{\alpha})) \exp(-\mu) \exp(-\mathbf{X}'\boldsymbol{\alpha}),$$

$$(3.5)$$

where λ_{A0} denotes the baseline hazard arising from the accelerated failure time representation. Since this equation holds for any covariate vector $\mathbf{X}$, we can choose particular values to get the results we need: At first, we set $\mathbf{X} = (-\mu/\alpha_1, 0, ..., 0)$ in (3.5). This yields

$$\lambda_{R0}(t) = \exp\left(\frac{\mu\beta_1}{\alpha_1}\right) \lambda_{A0}(t).$$

Secondly, we take $\mathbf{X}$ equal to $(ln(t)/\alpha_1, 0, 0, ..., 0)'$. Thus, we get

$$\lambda_{R0}(t) \exp\left(ln(t)\frac{\beta_1}{\alpha_1}\right) = \lambda_{A0}(t \exp(-ln(t)) \exp(-\mu)) \exp(-\mu - ln(t)).$$

Combining these two equations and simplifying the exp-log-parts yields

$$\lambda_{R0}(t) t^{\frac{\beta_1}{\alpha_1}} = \lambda_{R0}(\exp(-\mu)) \frac{1}{t} C,$$

where $C := \exp(-\mu)\exp(-\mu\beta_1/\alpha_1)$. So, we finally arrive at

$$\lambda_{R0}(t) = \lambda p t^{p-1},$$

where $\lambda := (-\alpha_1/\beta_1)C\lambda_{R0}(\exp(-\mu))$ and $p := -\beta_1/\alpha_1$. $\qquad\square$

Next, we turn to the discussion of parameter estimation. For this purpose, we will use the well-known maximum likelihood method. However, since we have to account for censoring, we have to examine what the likelihood function looks like before we can actually go on with the standard estimation procedure.

To set the stage, let $((t_i, \delta_i, \mathbf{x_i}))_{i=1}^n$ be realizations of a sample of censored survival times T_i with associated covariate vectors $\mathbf{X_i} := (X_{i,1}, X_{i,2}, ..., X_{i,p})'$, $1 \leq i \leq n$, as stated in Definition 2.6. Then, the following proposition holds (Klein and Moeschberger [16], pp. 76-77):

Proposition 3.5 (Likelihood for random censoring). *With the assumptions from above, the following statement holds: If the censoring is noninformative, we have*

$$L(\boldsymbol{\theta}) \propto \prod_{i=1}^n f(t_i|\boldsymbol{\theta}, \mathbf{x_i})^{\delta_i} S(t_i|\boldsymbol{\theta}, \mathbf{x_i})^{1-\delta_i} \tag{3.6}$$

Proof. The assumption of noninformative censoring implies that the density g_i and the cumulative distribution function G_i corresponding to C_i don't involve the parameter of interest $\boldsymbol{\theta}$. Thus, we write $g_i(t)$ and $G_i(t)$, $1 \leq i \leq n$.
The only thing we have to do is to take a look at the following distinct cases: If the i-th observation is a censored one, we know that $T_i = min(T_i^{ex}, C_i) = C_i$. Thus, we get

$$P(T_i = t, \delta_i = 0|\boldsymbol{\theta}, \mathbf{x_i}) = P(C_i = t, T_i^{ex} > t|\boldsymbol{\theta}, \mathbf{x_i})$$

$$= \frac{d}{dt} \int_0^t \int_v^\infty f(u|\boldsymbol{\theta}, \mathbf{x_i})g_i(v)du\,dv$$

$$= \frac{d}{dt} \int_0^t S(v|\boldsymbol{\theta}, \mathbf{x_i}) g_i(v) dv$$

$$= S(t|\boldsymbol{\theta}, \mathbf{x_i}) g_i(t),$$

for any $t \geq 0$. Note that we used the random censoring assumption, that is, the independence of C_i and T_i^{ex}, in the second step. Analogously, we handle the case where the i-th observation is un-censored (i.e., $T_i = min(T_i^{ex}, C_i) = T_i^{ex}$):

$$P(T_i = t, \delta_i = 1|\boldsymbol{\theta}, \mathbf{x_i}) = P(T_i^{ex} = t, C_i > t|\boldsymbol{\theta}, \mathbf{x_i})$$

$$= \frac{d}{dt} \int_0^t \int_v^\infty f(v|\boldsymbol{\theta}, \mathbf{x_i}) g_i(u) du\, dv$$

$$= \frac{d}{dt} \int_0^t f(v|\boldsymbol{\theta}, \mathbf{x_i})(1 - G_i(v)) dv$$

$$= f(t|\boldsymbol{\theta}, \mathbf{x_i})(1 - G_i(t)),$$

where G_i denotes the CDF of C_i. So, all in all, we arrive at

$$L(\boldsymbol{\theta}) = \prod_{i=1}^n P(T_i = t_i, \delta_i = 1|\boldsymbol{\theta}, \mathbf{x_i})^{\delta_i} P(T_i = t_i, \delta_i = 0|\boldsymbol{\theta}, \mathbf{x_i})^{1-\delta_i}$$

$$= \prod_{i=1}^n (f(t_i|\boldsymbol{\theta}, \mathbf{x_i})(1 - G_i(t_i)))^{\delta_i} (S(t_i|\boldsymbol{\theta}, \mathbf{x_i}) g_i(t_i))^{1-\delta_i}$$

$$= \prod_{i=1}^n (1 - G_i(t_i))^{\delta_i} g_i(t_i)^{1-\delta_i} \prod_{i=1}^n f(t_i|\boldsymbol{\theta}, \mathbf{x_i})^{\delta_i} S(t_i|\boldsymbol{\theta}, \mathbf{x_i})^{1-\delta_i}.$$

Due to the fact that the first product is a constant with respect to $\boldsymbol{\theta}$, we are done. $\square$

Remark 3.6. It can be shown by using counting process theory that the likelihood given above is correct even under more general assumptions, namely if the underlying censoring scheme is independent (see Equation 2.1 in chapter 2). For details, we refer to Kalbfleisch and Prentice [14], p. 53-54.

Example 3.7 (Weibull model). Let's consider the accelerated failure time representation of the Weibull model, namely

$$Y = \mu + \mathbf{X}'\boldsymbol{\alpha} + \sigma W,$$

where $Y := ln(T)$. Since we assume that W follows the extreme value distribution, we have (see Example 3.3):

$$f_W(w) = exp(w - exp(w)),$$
$$S_W(w) = exp(-exp(w)).$$

As shown in Example 3.3, the density and survival function for Y are then given by

$$f_Y(y_i) = \frac{1}{\sigma}exp\left(\frac{y_i - \mathbf{x_i}'\boldsymbol{\alpha} - \mu}{\sigma} - exp\left(\frac{y_i - \mathbf{x_i}'\boldsymbol{\alpha} - \mu}{\sigma}\right)\right),$$
$$S_Y(y_i) = exp\left(-exp\left(\frac{y_i - \mathbf{x_i}'\boldsymbol{\alpha} - \mu}{\sigma}\right)\right)$$

for $i = 1, 2, ..., n$.

Now, we proceed in the usual way: We substitute these expressions in (3.6) and calculate the first partial derivatives of $ln(L(\boldsymbol{\alpha}, \mu, \sigma))$ with respect to each of the parameters. Then, we set these derivatives (the so-called *score statistics*) equal to 0 and solve the resulting system of equations by applying some numerical technique like the Newton-Raphson procedure.

To illustrate the theoretical considerations from above, we take a look at the epilepsy data again. Using the `survreg` function in the **survival** package, we fit a Weibull regression model with *gender* (female is coded as 0 and male as 1, respectively), *age at diagnosis*, *etiology* (coded as dummy variables, such that symptomatic $= (1,0)$, idiopatic $= (0,0)$ and cryptogenic $= (0,1)$) and *year of diagnosis* to the data. Note that the R output gives the parameter estimates of the accelerated failure time representation of the model. To get the corresponding values for

Table 3.1: Results of Weibull relative risk regression model fit to the
epilepsy dataset, with *gender*, *age at diagnosis*, *etiology* and
year of diagnosis as covariates.

	estimate	standard error
λ	1.135×10^{31}	1.900×10^{32}
p	1.096	0.059
gender	0.378	0.132
age at diagnosis	0.077	0.004
dummy1	0.359	0.381
dummy2	-0.314	0.413
year of diagnosis	-0.044	0.008

the relative risk model, we use the `ConvertWeibull` function in the
package **SurvRegCensCov** (Hubeaux and Rufibach [13]).

As can be seen in the summary table (Table 3.1), the shape parameter p
is greater than one, which means that the risk of dying is monotonically
increasing with time. Moreover, as we expect, the older the people are
at time of diagnosis, the larger the hazard is. The coefficient estimate
for *year of diagnosis* is negative, thus suggesting that the patients who
entered the study later have a better chance of surviving. As to the
categorical variables, men seem to do worse than women. Moreover,
patients with symptomatic epilepsy tend to have the highest risk of
dying, followed by idiopatic and, finally, cryptogenic etiologies.

To close this chapter, we stay with the relative risk models once
more. Before we examined a special parametric model, namely the
Weibull model, we had stated that we could give examples of several
instances of relative risk models by just specifying the baseline hazard
parametrically. But what if we leave h_0 unspecified? Interestingly,
even in this general setting, we can apply an estimation procedure
which yields similar results as the standard maximum likelihood
method, at least as far as asymptotic properties of the estimators
are concerned. We don't give an extensive account of this so-called

semiparametric approach here, but we outline at least some of the basic ideas. To begin with, we introduce some additional notation. For an arbitrary $t \geq 0$, the set containing the subjects at risk just prior to t is denoted by $R(t)$. Furthermore, in addition to the setting above, we apply some re-indexing to our data such that the k exact survival times out of n observations in total are denoted by $t_1 < t_2 < ... < t_k$, $k \leq n$ (see chapter 2). By the way, the assumption that there are no ties between the exact survival times can be relaxed without facing serious problems (see Klein and Moeschberger [16],pp. 259-260). Apart from these modifications, the assumptions remain the same as outlined at the beginning of the paragraph dealing with estimation in regression models. Especially, we consider a random censoring scheme.

Definition 3.8 (Relative risk partial likelihood). With the conventions from above, let us assume a relative risk model

$$\lambda(t|\mathbf{X}) = \lambda_0(t)\exp(\mathbf{X}'\boldsymbol{\beta})$$

for the data. Then, the *Relative risk partial likelihood* or *Cox partial likelihood* is defined by

$$\prod_{i=1}^{k} \frac{\exp(\mathbf{x_i}'\boldsymbol{\beta})}{\sum_{k \in R(t_i)} \exp(\mathbf{x_k}'\boldsymbol{\beta})}$$

Remark 3.9.

1. Actually, a so-called partial likelihood can be defined in a much more general probability theoretic setting. Then, the definition from above could be regarded as a proposition (Kalbfleisch and Prentice [14], pp. 99-102).

   Alternatively, the expression given in Definition 3.8 can also be derived as a profile likelihood of expression (3.6), i.e. the "full" likelihood function for censored data (Klein and Moeschberger [16], p. 258).

2. Keep in mind that the terminology must not be misinterpreted: The "likelihood" defined above is not a likelihood in the usual

Table 3.2: Results of `coxph` for the epilepsy data.

	estimate	standard error
gender	0.394	0.133
age at diagnosis	0.078	0.005
dummy1	0.371	0.380
dummy2	-0.306	0.414
year at diagnosis	-0.042	0.009

sense. But interestingly, this function can nevertheless be treated like a "normal" likelihood as far as estimation and asymptotic properties are concerned (Kalbfleisch and Prentice [14], p. 100).

3. In the more general setting of independent censoring, the partial likelihood can be defined similarly to Definition 3.8.

Due to remark (3.9)(2), we can now proceed in the standard way. That is, we differentiate the log-partial likelihood with respect to each of the components of β, set these partial derivatives equal to 0 and calculate the desired solutions by using some numerical procedure. Moreover, one can show that the resulting "maximum likelihood" estimators have asymptotic properties which are completely analogous to those for parametric models like the Weibull model. To illustrate this method, let's have a look at the epilepsy dataset once more.

Example 3.10. Analogously to the Weibull model example above (see Example 3.7), we fit a relative risk model to the data, using the `coxph` function in the **survival** package. The results are displayed in Table 3.2. To cut it short, the estimates and associated standard errors are very close to the corresponding values seen in Example 3.7. By the way, this somehow indicates that assuming a Weibull model for the data is quite a reasonable choice. This issue will be examined more thoroughly in the next chapter.

However, there's one additional task left to do: Remember that we've assumed a relative risk model for the data. But so far, we have only calculated estimates of the regression coefficients, but not of the baseline hazard. Interestingly, parameter estimation worked well without using any information about λ_0! When taking a look at the derivation of the partial likelihood defined above, one can see that originally, the baseline hazard is involved in the formula, but cancels out afterwards (Kalbfleisch and Prentice [14], p. 102). However, when we want to calculate a basic quantity such as the survival function, an estimate of the baseline (cumulative) hazard is apparently needed. For example, we could use the so-called *Breslow estimator* of the baseline cumulative hazard function (Breslow [2]), which is defined as

$$\hat{H}_0(t) := \sum_{\{i:t_i \leq t\}} \frac{1}{\sum_{k \in R(t_i)} exp(\mathbf{x_k}'\hat{\beta})},$$

where $\hat{\beta}$ denotes the vector containing the partial ML estimates. This estimator can be derived, for example, by a profile likelihood approach, as mentioned in remark 3.9. By using the results of Theorem 2.5, we get $\hat{S}_0 := exp(-\hat{H}_0)$ and, finally, an estimate of the survival function S because in a relative risk model, we have $S(t|\mathbf{X}) = S_0(t)^{exp(\mathbf{X}'\beta)}$ (this follows immediately from Definition 3.1 and the basic relations established in chapter 2).

It should be added that quite a broad variety of alternative estimators of the baseline survival function have been proposed in literature. For details concerning this issue as well as the properties and derivations of some of these estimators, see Klein and Moeschberger [16], pp. 283-287 and Kalbfleisch and Prentice [14], pp. 114-118. An illustration of some of these estimators is given in Example 5.15.

4 Model checking procedures

In this chapter, we will basically address two important issues: Firstly, we will present some methods for checking, let's say, the basic assumptions underlying a particular model. For example, when considering a Weibull model, we have to answer the following questions:

1. Does the proportional hazards assumption hold (see chapter 3)?

2. Is it reasonable to assume that the survival times follow a Weibull distribution?

Secondly, we will discuss a topic which is usually referred to as model building. Although the underlying problems are the same as in linear regression models, the survival analytic setting requires different methods to calculate residuals etc., as we will see in the second part of this chapter. In addition to the books of Klein and Moeschberger [16], pp. 353-364, 409-419, and Kalbfleisch and Prentice [14], pp. 119-128, 210-212, we will at some points use other references, too, as cited in the text.

To begin with, let's recall what the proportional hazards assumption means: Suppose we have specified a semiparametric or parametric regression model by assuming that the hazard is given as follows:

$$\lambda(t, \mathbf{X}) = \lambda_0(t) \exp(\mathbf{X}'\boldsymbol{\beta}).$$

Now, if we assume that the covariates (i.e., the components of $\mathbf{X}$) are time-independent, the so-called **proportional hazard assumption** holds, that is, if we take two arbitrarily chosen individuals with covariate vectors $\mathbf{x_1}$ and $\mathbf{x_2}$, we then have

$$\frac{\lambda(t|\mathbf{x_1})}{\lambda(t|\mathbf{x_2})} = \frac{\lambda_0(t) \exp(\mathbf{x_1}'\boldsymbol{\beta})}{\lambda_0(t) \exp(\mathbf{x_2}'\boldsymbol{\beta})} = \frac{\exp(\mathbf{x_1}'\boldsymbol{\beta})}{\exp(\mathbf{x_2}'\boldsymbol{\beta})}$$

for every $t \geq 0$.

Equivalently speaking, if we set

$$C_{\mathbf{x_1},\mathbf{x_2}} := \frac{\exp(\mathbf{x_1}'\beta)}{\exp(\mathbf{x_2}'\beta)},$$

we get

$$\frac{\Lambda(t|\mathbf{x_1})}{\Lambda(t|\mathbf{x_2})} = \left(\int_0^t \lambda(s|\mathbf{x_1})ds\right)\left(\int_0^t \lambda(s|\mathbf{x_2})ds\right)^{-1}$$

$$= \left(C_{\mathbf{x_1},\mathbf{x_2}}\int_0^t \lambda(s|\mathbf{x_2})ds\right)\left(\int_0^t \lambda(s|\mathbf{x_2})ds\right)^{-1}$$

$$= C_{\mathbf{x_1},\mathbf{x_2}}.$$

Eventually, using Theorem 2.5, we arrive at

$$ln(-ln(S(t|\mathbf{x_1}))) - ln(-ln(S(t|\mathbf{x_2}))) = ln\left(\frac{\Lambda(t|\mathbf{x_1})}{\Lambda(t|\mathbf{x_2})}\right) = ln(C_{\mathbf{x_1},\mathbf{x_2}}).$$

So, to check if the proportional hazards assumption is fulfilled, we calculate an estimate $\hat{S}_{\mathbf{x}}$ (e.g., the Kaplan Meier estimate) for each possible covariate vector $\mathbf{x}$, apply the $ln(-ln)$ transformation to these estimates and plot the resulting values against t. If the lines look parallel, we may assume that the proportional hazards assumption is not violated. Note that the distance between the lines need not necessarily be the same for all covariate vectors, as indicated by the subscript $\mathbf{x_1},\mathbf{x_2}$ above.

When dealing with parametric models, the question naturally arises if the chosen specific function fits the data reasonably well. To examine this issue, we seek a transformation of the survival function S which is linear in some function of time t. For example, the baseline survival function of a Weibull model is given as

$$S_0(t) = \exp(-\lambda t^p).$$

Now, just as before, we apply a $ln(-ln)$ transformation and get

$$ln(-ln(S_0(t))) = ln(\lambda) + ln(t)p.$$

Thus, it turns out that $ln(-ln(S_0(t)))$ is linear in $ln(t)$. If we consider a Weibull regression model, this basic result does not change substantially since we then have

$$ln(-ln(S(t|\mathbf{x}))) = ln(\lambda) + \mathbf{x}'\boldsymbol{\beta} + ln(t)p. \qquad (4.1)$$

Therefore, to check the appropriateness of a Weibull regression model, we calculate Kaplan Meier estimates for each possible subgroup of the data. Note that in contrast to the PH assumption check carried out before, we now plot these estimates against $ln(t)$. If the resulting lines look linear, the model fits the data reasonably well.

At least for the Weibull model, the means of model checking just introduced can be used for the examination of the PH assumption, too. To see this, let's take a look at equation (4.1) once more. Obviously, the shape parameter p of the Weibull distribution corresponds to the slope of $ln(-ln(S(t|\mathbf{x})))$. Why's this important? Well, for the validity of the PH assumption, the shape p is required to be one and the same value for all possible covariate combinations: If we assume that $p_1 \neq p_2$ for some covariate vectors $\mathbf{x_1}, \mathbf{x_2}$, we get

$$\frac{\lambda(t|\mathbf{x_1})}{\lambda(t|\mathbf{x_2})} = \frac{\lambda p_1 t^{p_1-1} \exp(\mathbf{x_1}'\boldsymbol{\beta})}{\lambda p_2 t^{p_2-1} \exp(\mathbf{x_2}'\boldsymbol{\beta})}.$$

Obviously, the expression on the right depends on t (note that $p_1 \neq p_2$ implies $p_1 \neq 1 \vee p_2 \neq 1$!), which means that the PH assumption is violated.

So, to sum things up, all we have to do is to calculate estimates $\hat{S}_{\mathbf{x}}$ for every possible covariate vector $\mathbf{x}$. Then, there are basically two means of checking proportional hazards:

1. In a **relative risk model with time-independent covariates**, we can check the proportional hazards assumption by plotting $ln(-ln)(\hat{S}_{\mathbf{x}})$ against **time t**. If the resulting graphs look **parallel**, we may assume that the proportional hazards assumption holds.

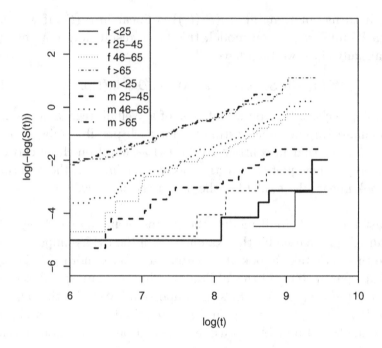

Figure 4.1: PH and Weibull assumption check for the epilepsy data

2. If we decide to fit a parametric regression model such as the **Weibull regression model** to the data, we should, at first, try to find a suitable transformation such as $ln(-ln)$. Then, we plot $ln(-ln)(\hat{S}_\mathbf{x})$ against $ln(t)$, the **ln of time**. Like in the previous case, **parallelity** (i.e., equal slopes) of the graphs indicates that the **hazards are proportional**. Moreover, **straight lines** support the assumption of a **Weibull model** for the data.

Example 4.1. To examine the Weibull model assumption for the epilepsy data, let's divide the dataset into subgroups according to *gender* and the four age groups < 25, 25-45, 46-65 and >65, as in Example 2.21. In that example, we've already noticed that the number of observed exact survival times is very small for some subgroups. Therefore, it's hard to decide if the proportional hazards assumption is fulfilled. If you look at the $ln(-ln)$ transformations of the KM estimates plotted against $ln(t)$, as displayed in Figure 4.1, it seems reasonable to assume proportional hazards for male and female patients between 46 and 65 and over 65, respectively, since the curves seem to be parallel. In addition to that, the lines look straight, which supports the assumption of a Weibull distribution.

But, what can you tell about parallelity or linearity for the first two age groups? In fact, hardly anything or - for those younger than 25 - nothing. But, now note that we have only accounted for two covariates so far. Thus, if we also take, for example, epilepsy etiology, into consideration, things will get even worse since every subgroup is split up into (potentially) three smaller ones.

Apart from this problem, it may be not easy to judge if the lines made up by the KM estimates are linear and/or parallel because they are step functions. Thus, this method is only of limited use for gathering evidence that the model assumptions hold. The KM plots should rather be used to detect severe violations such as nonproportionality of hazards, which corresponds to crossing lines seen in the plot.

As we have learned in the previous example, the method of plotting some suitable transformation of the Kaplan Meier estimates against some function of time is quite easy to carry out. Nevertheless, especially when dealing with small sample (or subgroup) sizes, this means of checking doesn't work well any more. Therefore, we have to seek for alternative procedures. For example, to examine the overall fit of a parametric model, we can use certain residuals: The so-called

Cox-Snell residuals are defined as

$$r_j := \hat{\Lambda}(t_j) = \hat{\Lambda}_0(t_j) \exp(\mathbf{x_j}'\hat{\boldsymbol{\beta}}), j = 1, 2, ..., n,$$

where $\hat{\Lambda}$ and $\hat{\boldsymbol{\beta}}$ are the estimates of the cumulative baseline hazard and the regression coefficients, respectively. The following proposition basically contains the main idea how to perform a goodness-of-fit check based on these residuals.

Proposition 4.2. *Let T be a non-negative continuous random variable. Then, we have*

$$\Lambda(T) \sim \exp(1).$$

Proof. Since T is continuous, we have that $Y := F_T(T) \sim U(0,1)$: For any y in $(0,1)$,

$$F_Y(y) = P(Y \leq y) = P(F_T(T) \leq y) = P(T \leq F_T^{-1}(y))$$
$$= F_T(F_T^{-1}(y)) = y.$$

Moreover, if we let $U := \Lambda(T)$, it follows from Theorem 2.5 that

$$F_U(t) = P(U \leq t) = P(-ln(S(T)) \leq t) = P(-ln(1 - Y) \leq t)$$
$$= P(Y \leq 1 - \exp(-t)) = F_Y(1 - \exp(-t))$$
$$= 1 - \exp(-t).$$

Because probability distributions are uniquely determined by their CDFs, we are done. $\square$

As to the interpretation of the statement just proven, Proposition 4.2 says that the Cox-Snell residuals should look like a censored sample from an exponential distribution with scale parameter 1 (note that introducing covariates wouldn't affect the calculations carried out above!). Consequently, our goal is to calculate an estimate of a suitable survival analytic statistic for the residuals (!) and compare the result with the corresponding quantity of an exp(1) variable. To make interpretations as easy as possible, we decide to take the cumulative hazard function since for $X \sim \exp(1)$, we have $\Lambda_X(x) = x$. As to the

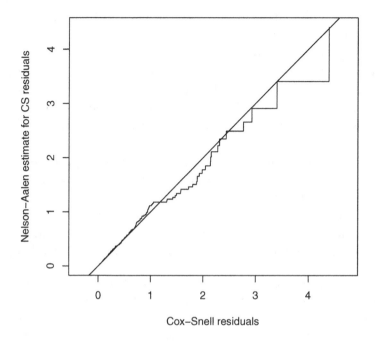

Figure 4.2: Cox-Snell residuals for the epilepsy data

residuals, we use the Nelson-Aalen estimate of the cumulative hazard function (see chapter 2). We illustrate this means of assumption checking in the following example.

Example 4.3. Let's have a look at the Cox-Snell residuals for the epilepsy data. We assume a Weibull model for the data, that is, the Cox-Snell residuals are calculated as

$$r_j = \hat{\lambda}\hat{p}t^{\hat{p}-1}\exp(\mathbf{x_j}'\hat{\boldsymbol{\beta}}), j = 1, 2, ..., n,$$

where we used the maximum likelihood estimates from the example in the previous chapter. The plot of the Nelson-Aaalen estimate of

the residuals' cumulative hazard vs. the values of the residuals is displayed in Figure 4.2.
For most of the data points (recall that each step corresponds to an exact death time!), the Nelson-Aalen estimate is very close to the reference line $\Lambda_X(x) = x$. Note that we don't have to worry about the large deviations at the far right because the variance of the Nelson-Aalen estimator is known to increase when the "survival times" get larger.
However, it must be mentioned that this way of assumption checking only gives us very general information about the appropriateness of the model we've chosen. In addition to that, it has been noticed that the estimated cumulative hazard of the residuals may look like the exp(1) cumulative hazard, although the fit is quite poor (for some references concerning this issue, see Kalbfleisch and Prentice [14]).

So, summing up, we've seen that with slightly more theoretical efforts than before, we can find a means of model checking, namely the Cox-Snell residual plot, which is less troubled by a high degree of censoring or small subgroup sizes - note that we didn't divide the data into subgroups in the example above! - than the comparison of Kaplan Meier estimates. However, even with this method, small sample sizes may still cause difficulties. Moreover, we only get a very rough idea of the model's goodness-of-fit (see Kalbfleisch and Prentice [14], p. 213 for further information).

To close this part of the chapter, we briefly discuss a formal test which can be used to check the proportional hazards assumption. We won't go into details since it would take too much time to introduce the concepts from martingale and counting process theory needed to derive the test formally. But we will give at least the main idea how the test works. For the technical details, we refer to Grambsch and Therneau [9]. Until now, we have only dealt with time-independent impact of the covariates. But, it is also possible to account for time-dependency in a relative risk model. Of course, the hazards wouldn't be proportional to each other any more, then. Thus, if we want to

Table 4.1: Results of `cox.zph` for the epilepsy data.

	correlation	test statistic	p value
gender	0.0258	0.1593	0.690
age at diagnosis	-0.0661	0.9541	0.329
dummy1	0.0368	0.3249	0.569
dummy2	0.0177	0.0771	0.781
year of diagnosis	0.0606	0.7236	0.395
global	NA	2.2870	0.808

examine the proportional hazards assumption, we just replace the j−th regression coefficient β_j by $\beta_j(t) := \beta_j + \theta_j g_j(t)$, where g_j is some specified function of time t, and try to find a test for $H_0 : \theta_j = 0$. Grambsch and Therneau [9] derive an asymptotic χ^2 test for this problem, where the test statistic is based on certain residuals. This test can be carried out using the function `cox.zph` in the **survival** package in R (Therneau [26]), as demonstrated now.

Example 4.4. To examine the proportional hazards assumption for the epilepsy dataset, we fit a Cox model with *gender*, *age at diagnosis*, *epilepsy etiology* (coded as dummies) and *year of diagnosis* to the data using the `coxph` function.
Then, we call `cox.zph` with the default values for the arguments and have a look at the resulting table (see Table 4.1). Note that the NA appears due to the fact that the `cox.zph` function doesn't provide a correlation coefficient for the global hypothesis. Obviously, there isn't any evidence suggesting that the PH assumptions is violated. However, this doesn't mean that the hazards are indeed proportional: Although this means of proportional hazards checking doesn't require a division of the patient cohort into several subgroups, we have to be aware of the fact that the derivation of the tests carried out above relies on asymptotic results, which might be troubled by a relatively high degree of censoring. In addition to that, due to construction (see above), the power of the test may heavily depend on the choice of the functions g_j.

In this context, it is worth mentioning that there are alternative tests for checking the proportionality of the hazards available: For example, we can as well use a method based on a likelihood ratio test. This issue will be touched again in the variable selection part of this chapter.

To sum things up, we have discussed several means of checking the basic assumptions of a survival analytic model. The main point we have found is that a widely used procedure such as the check based on the Kaplan-Meier estimates can be seriously affected by a high degree of censoring and/or small subgroup sizes. Nevertheless, the alternatives we have discussed also suffer from some drawbacks. Consequently, we suggest using not only one single method, but several means of model checking. Thus, hopefully, we will be able to reduce the uncertainities associated with each single method and eventually get a fairly reliable answer to the question whether the basic model assumptions are met or not.

By the way, although scientific progress will likely be made in future years, we have to be aware of the fact that for certain situations, it will hardly be possible to find substantial improvements. To illustrate this aspect, just think of a hypothetical dataset where, let's say, all male patients are censored. Now, if you are interested in the survival experience, what can you tell about the survival times of these persons? According to the discussions in previous chapters, this kind of data only allows moreorless rough statements, for example, "their mean survival time is at least 3.4 years". But, what about their exact survival times? Nobody can answer this question based merely on the information available. Now, it should be clear that likewise, we can only make very rough statements about the proportionality of hazards with respect to, let's say, gender. To cut it short, the problem of finding a proper method for assumption checking might not only depend on your statistical genius, but also on the information the data provides.

Let us now turn to the second question mentioned in the introductory part of this chapter, namely, to some issues concerning model building and variable selection. To begin with, let's recall which type of data requires some thoughts about this topic. For example, in a randomized clinical trial where the main point of interest concerns the question whether a certain drug makes, for instance, the hazard rate decrease or not, we usually don't have to worry about the in- or exclusion of variables despite, of course, the group indicator: If randomization worked well, the characteristics of the groups which are compared should be quite similar to each other.

However, if you retrospectively analyze data on the survival experience of some patients, you definitely have to take several covariates into account: For example, the survival times most likely depend on the subject's age and gender. Apart from these simple considerations, some medical experts can tell you which additional variables should be incorporated in the model. In practice, the number of variables often tends to get higher and higher quickly: Usually, the collection and preparation of data for biostatistical analyses is quite time-consuming and may cost a lot of money. So, the researchers sometimes try to present as much information as possible in their publications. Moreover, if you merely consider some very basic thoughts, the number of variables that should be included may increase substantially: We have already mentioned that usually, age and gender of the patients should be taken into consideration. If the data is collected over several years, the year of entry into the study or something like that should be included, too. Then, if we assume that we are interested in, let's say, the impact of cancer on survival, it's clear that we should distinguish between lung cancer, kidney cancer and so on. Furthermore, the survival experience most likely depends on the development of metastases. So, moreorless with referring to common sense only, we have already identified five variables which should be included in the model. For sure, an oncologist would rightfully advocate for taking even more variables into account due to medical reasons.

However, common sense and the opinion of a subject matter expert may be not enough to ensure that we have a good chance to get reliable results. For example, it is known that in regression analysis, adding more and more variables to a model may eventually lead to very high variances of the regression coefficient estimates. Moreover, the Wald tests for significance of the regression coefficients can become fairly conservative. The reason for these problems is a phenomenon called multicollinearity, that is, at least some of the variables may be highly correlated. So, all in all, we have to find a balance between a number of variables which is too small (as mentioned before, this may cause severe bias) and a model that is "overfitted". Although it's difficult to state a rule of thumb for this issue, we can study the behaviour of the regression coefficients etc. by conducting simulations, such as Cocato et al. [4] and Peduzzi et al. [21], who suggest that the number of events (i.e., deaths) per variable should be at least 10.

In addition to that, like in linear and logistic regression, there are backward and forward variable selection methods availabe for survival analytic models, too (Klein and Moeschberger [16], pp. 276-282 Collett [5], pp. 80-87). They are based, for example, on the Akaikie information criterion (AIC), which is defined as

$$AIC = 2 * (p + k - LL),$$

for (semi-)parametric survival analytic models, where p is the number of variables, LL denotes the log likelihood evaluated at the maximum likelihood estimates of the regression coefficients and k is a model-specific predetermined constant (e.g., $k = 0$ for the Cox model and $k = 2$ for the Weibull model, respectively). Now, the main goal is to get small values of AIC. So, for example, if we use the AIC for variable selection within a certain family of models, only p and LL can vary. Ideally, the latter quantity should increase by more than 1 if we add an additional variable to the model in order to arrive at a lower AIC. By the way, as the AIC also involves a model-specific parameter k, this criterion can be used for comparisons between models belonging to different parametric families, too. However, we don't want to go into details and refer to Klein and Moeschberger [16], pp. 406-407, for further information about this topic.

By the way, it should be mentioned that variable selection can also be carried out by using *(sequential) local tests* for significance of the regression coefficients. More to the point, if we, for example, want to examine whether a certain additional variable should be incorporated in the model, we can use a likelihood ratio test: It can be shown that under the null hypothesis of no difference between the "basic" and the "extended" model, the test statistic

$$LR := 2 * (LL_e - LL_b)$$

is asymptotically chi-squared distributed with one degree of freedom, where LL_e and LL_b denote the log likelihood of the extended and the basic model, respectively. For a derivation, see Klein and Moeschberger [16], pp. 449-452. Note that in contrast to the AIC, this method can be used for variable selection only, or, equivalently speaking, for nested models, but not for comparisons between models from different parametric families. Nevertheless, local tests like the one proposed above can be very useful in a fairly broad range of problems. For example, they can also serve as a tool for checking the proportional hazards assumption, or, in other words, for modelling a time-varying influence of a particular covariate: Just test for significance of an additional interaction term between that certain covariate and time (or some function of time).

Furthermore, we can as well use the likelihood ratio test for examining interactions between two covariates, as demonstrated in the following example.

Example 4.5. Due to the fact that epilepsy etiology may depend on the age of the patient, it is reasonable to examine whether an interaction term between *age at diagnosis* and *etiology* should be included in the model or not. Therefore, we fit two Weibull regression models, one of them with *gender, age at diagnosis, epilepsy etiology* and *year of diagnosis* as covariates (see chapter 3), the second one with the additional interaction term *age at diagnosis* × *etiology*. Recall that in a software package like R, we code *etiology* by using two dummy variables *dummy1* and *dummy2*. Thus, we actually add two variables to the original model, namely age × *dummy1* and age × *dummy2*.

The AIC for the first model is 4933.12, whereas for the latter one, we get a value of 4937.10. This very small difference seen in the AICs suggests that the interaction between age and etiology only has a negligible influence on the survival experience of the cohort. This can be also seen by conducting a likelihood ratio test, as proposed above, which gives a p-value of 0.989. Note that for the latter test, we need a generalization of the result from above: Informally speaking, if we add k variables to a specific regression model, it can be shown that two times the difference of the log likelihoods is asymptotically chi-squared distributed with k degrees of freedom (Klein and Moeschberger [16], p. 264). So, in our example, we compare our test statistic with the appropriate quantile of a chi-squared distribution with 2 degrees of freedom.

In the previous example, we have seen that the methods of variable selection aren't restricted to the question which variables we should pick from our dataset. To use the terminology of Kalbfleisch and Prentice [14], p. 96, there may be also "derived" covariates such as interaction terms or transformations of the original variables. Let us briefly discuss the latter issue now. Suppose that we are interested in finding a function f such that the incorporation of $f(X_i)$ in our model fits the data reasonably well, where X_i is a certain covariate. Of course, we could use formal statistical tests, as outlined above. However, in addition to that, it's recommended to perform some graphical checks, too (Klein and Moeschberger [16], p. 353, Collett [5], pp. 111, 148-149). If we had assumed a linear model for the data, this would have been a quite easy task: Just plot the residuals of the basic model, say, $y_i - \hat{y}_i$, against the values of x_i. If you additionally let R draw some smoothed line, you will get at least a rough idea what f

should look like. But, when dealing with survival analytic settings, things are more complicated. Nevertheless, under the restrictions of right-censoring and time-independent covariates, it can be shown that, analogously to the residuals in linear models, the quantity

$$\hat{M}_j := \delta_j - r_j, 1 \leq j \leq n,$$

can be interpreted as the difference "observed minus expected values". Recall that r_j and δ_j are the Cox-Snell residual and the censoring indicator for the j-th subject, respectively.

By the way, the reason for the hat-notation is that if we replace the sample values used for the calculation of the quantities on the right handside by the true values, the resulting expression turns out to be a martingale. Therefore, the $\hat{M}_j$ are usually referred to as *martingale residuals*. For theoretical details, we refer to Klein and Moeschberger [16], p. 362.

Example 4.6. To illustrate the method proposed above, let's assume that we are interested in determining the appropriate functional form for the covariate *age at diagnosis*. At first, we fit a Weibull regression model with *gender* and *etiology* to the epilepsy data. Note that the data meets the assumptions mentioned above: We only have to account for right censoring and time-independent covariates!
Then, like in Example 4.3, we calculate the Cox-Snell residuals and plot the difference between the censoring indicators and these residuals against the values of *age at diagnosis*. To get at least a rough idea what the functional form of this covariate should look like, we additionally call the **lowess** function in R, which computes a smoothed curve using locally-weighted polynomial regression. The resulting plot is displayed in Figure 4.3.

At first sight, it seems as if the points could be divided into two groups. To examine this issue, we calculate the residuals for subsets according to the gender of the patients. But the results show that the two data

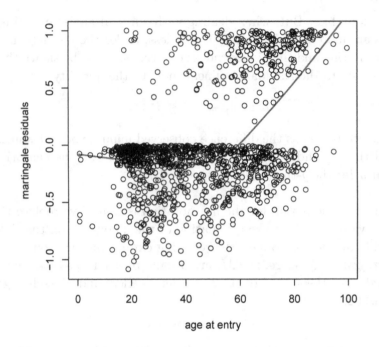

Figure 4.3: Martingale residuals to examine the appropriate functional
form for *age at diagnosis* in a Weibull regression model fitted
to the epilepsy dataset.

clouds don't represent the martingale residuals of males and females,
respectively. Likewise, the separation doesn't occur due to different
etiologies or a combination of etiology and gender. So, summing up,
the pattern seen in the plot can't be explained by referring to the
(categorical) covariates available. However, it might be possible that
another variable, which isn't accounted for, causes the separation.

Next, we continue with examining some kind of an overall trend. The plot indicates that up to an age of about 50, there seems to be hardly any impact on survival, whereas from 50 onwards, we can see an almost linear trend. So, we may think of f as a piecewise linear function. However, we should try to maintain some balance between choosing f such that the functional form fits the data well and the interpretability of the transformation. As to our example, it may be reasonable to leave the covariate unchanged, that is, to assume a linearly increasing impact on survival for the ages up to 50, too. To evaluate the appropriateness of certain transformations, we could, for example, compare several models with different choices of f by taking a look at the corresponding AIC values. So, again, we see that usually, more than one checking procedure is needed to get reasonable answers to our questions of model building.

5 Life expectancy

The main goal of this chapter is to discuss a method for comparing life expectancies, as proposed in Zimmermann [31]. To begin with, let us recall some facts about the interpretation of several survival analytic key quantities. Especially, I'd like to emphasize once again that the survival function S basically provides information on surviving beyond a specified time point t, whereas the hazard rate can be interpreted as a measure of the instantaneous risk of experiencing the event at time t. These quantities are often very useful, particularly when they are not only calculated for the cohort as a whole, but also for subsets corresponding to different age groups etc. If you consider a clinical study, the doctors are thus able to give very specific information about the survival of patients with certain characteristics. However, although the measures mentioned above are not hard to interpret in the scientific context, some patients might not be satisfied with such a kind of information: For example, imagine you are told that you are three times more likely to die than a comparable person from some reference population. Naturally, you would ask: "What does this mean? Will I die in one year? Or is there some chance to survive nearly as long as some healthy guy?" In other words, you would be very interested in a somehow more concrete information on your future survival experience. Apparently, this is an important question we have not touched so far. So, we need to introduce some new concepts now. At first, let us formulate the problem outlined above in a more mathematical way. To consider a general setting, assume that a certain time span t has passed since the start of the study. What the (hypothetical) patient in our example wants to know is the expected remaining lifetime. The word "expected" indicates which statistical quantity is chosen to answer the patient's question (see Kalbfleisch and Prentice [14], pp. 7-8, for the following definition as well as for the associated results presented below):

Definition 5.1 (Mean (residual) life at time t). Let T be a nonnegative continuous random variable with $E[T] < \infty$. Let $t \geq 0$. The *mean (residual) life at time t* or *expected residual life at time t* is defined as

$$r(t) := E[T - t | T \geq t].$$

The following proposition shows how $r(t)$ can be calculated.

Proposition 5.2. *For all $t \geq 0$, the following equation holds:*

$$r(t) = \frac{1}{S(t)} \int_t^\infty S(u)du.$$

Proof. Since we assumed in the definition of the mean residual life that $E[T] < \infty$, we have

$$\lim_{x \to \infty} xS(x) = 0. \tag{5.1}$$

To prove this, first note that the following line holds for all $x \geq 0$:

$$0 \leq x(1 - F(x)) = x \int_x^\infty f(y)dy = \int_x^\infty xf(y)dy \leq \int_x^\infty yf(y)dy.$$

Now, as x goes to infinity, the expression on the right side converges to 0 because $E[T] < \infty$.

To turn back to the statement we actually have to prove, we take the definition of the r(t), integrate by parts and finally apply (5.1):

$$r(t) = E[T - t | T \geq t] = \int_t^\infty \frac{(u - t)f(u)}{S(t)}du$$

$$= \frac{1}{S(t)} \lim_{x \to \infty} \left(\int_t^x (u - t)f(u)du \right)$$

$$= \frac{1}{S(t)} \lim_{x \to \infty} \left(-(u - t)S(u)|_t^x + \int_t^x S(u)du \right)$$

$$= \frac{1}{S(t)} \left(0 + \int_t^\infty S(u)du \right)$$

This completes the proof.

$\square$

Remark 5.3.

(i) The formula above suggests that the mean residual life is closely related to the other survival analytic quantities established previously. Indeed, it can be shown that the mean residual life uniquely determines the distribution of a nonnegative continuous random variable T. For details, we refer to Klein and Moeschberger [16], pp. 21, 35.

(ii) If we set $t = 0$, we get

$$\mu := E[T|T \geq 0] = E[T] = \int_0^\infty S(u)\mathrm{d}u,$$

which is called the *mean life* or *expected life*.

(iii) In many applications, it makes sense to replace the upper limit ∞ by some "cutoff value" t_{max}. For example, when death is the event of interest, it would not be realistic to assume that one can survive beyond a certain age. We will touch this idea again in the context of life table calculations.

Next, we turn to a different concept of life expectancy calculations, which is popular in the field of official statistics. To understand this approach in detail, we describe how the main quantities in a so-called population life table are calculated (Hanika and Trimmel [11]).

To begin with, we have to distinguish between two types of life tables according to the subject groups the tables provide survival information on: A *cohort life table* is based on the survival experience of a certain group of people with a common time origin (e.g., all people who were born in 1980), whereas a *population* or *period life table* contains information about the whole population (e.g., the population of Austria) for a certain year.[1] It should be emphasized that there is a big

---

[1] Sometimes, the time period stretches over, let's say, three years in order to avoid that, for example, a severe influenza epidemic has an undue influence on the resulting life table.

difference concerning the underlying information: When calculating a cohort life table, one uses data from the (common) time origin up to the death of the last surviving member of the initial cohort, whereas for a population life table, one, in general, collects information only within a certain year. To further elucidate this, just consider a simple example: It is not yet possible to calculate a cohort life table for the newborns in Austria in 2014 since most likely, some of them are still alive. However, we are able to set up an Austrian population life table for 2014 since all we need is data on the age- and gender-specific number of deaths in this certain year.

Despite the important difference outlined above, the general outlook of those two types is quite similar: Both provide death probabilities and related quantities for various age groups as well as male and female. As we will only use the population life table methodology in the following chapters, we do not further discuss cohort life tables here. For some basic information about the latter topic, we refer to Klein and Moeschberger [16], pp. 152-158, and Kalbfleisch and Prentice [14], pp. 19-20.

To begin with, the first column of a population life table contains the *exact ages* $x = 0, 1, 2, ..., x_{max}$. By "exact age", we mean the age at the x-th birthday. Furthermore, the value of x_{max} has to be chosen appropriately, e.g. $x_{max} = 95$ or $x_{max} = 100$.
Next, we give the definitions of several key quantities of a population life table.

Definition 5.4 (Population life table quantities). Let T denote the age at death (i.e., the survival time measured from birth as starting point). Let $x \in \{0, 1, ..., x_{max}\}$ be the exact age.

(i) The *death probability in the age interval* $[x, x + 1)$ is defined as

$$q_x := P(x \leq T < x + 1 | T \geq x).$$

(ii) Let $l_0 \in \mathbb{N}$. The *number of survivors at age* x is defined as

$$l_x := \begin{cases} l_0 & x = 0 \\ l_{x-1} \cdot (1 - q_{x-1}) & x \geq 1 \end{cases}$$

(iii) The *number of person years lived in the age interval* $[x, x+1)$ is defined as

$$L_x := \frac{1}{2} (l_x + l_{x+1}).$$

Remark 5.5.

ad(i): For calculating q_x based on actual data, several methods are available. Furthermore, one has to pay attention to the death probabilities of newborns ($x = 0$) and very old people ($x = x_{max}$) because special methods are needed to get the desired quantities for these age groups. But as we will either look at those quantities from a probability theoretic point of view or, especially in the real-data examples, treat them as given values, we do not discuss the actual calculation issues here and refer to Hanika and Trimmel [11] for some details.

ad(ii): In order to understand the main idea behind that definition, think of l_0 representing the initial size of some hypothetical (!) population (e.g., $l_0 = 100000$ is a commonly chosen value). Next, we assume this population is exposed to the risk of dying according to the death probabilities q_x calculated before. So, one should always keep in mind that the amount of "survivors" at a certain age doesn't refer to the situation in any real-life population! By the way, we see a very important point here: In fact, the death probabilities turn out to be the essential quantities of a population life table insofar as they already contain the whole information about the survival experience of the actual population. Once they are calculated, we don't need the data any more because all the other life table quantities mentioned above can be expressed in terms of the death probabilities.

ad(iii): The quantity L_x represents the number of years lived by the survivors aged x in the interval $[x, x + 1)$, that is, from their x-th to their $(x + 1)$-th birthday. It is clear that those people who survive beyond such an interval contribute l_{x+1} to L_x. But we have to take into account that, say, d_x out of the l_x survivors aged x die in this interval. More to the point, we know the exact number of those deaths: $d_x = l_x - l_{x+1}$. If we assume the deaths to be uniformly distributed in the interval $[x, x + 1)$, the people dying in that interval contribute $d_x/2$ to the quantity L_x. To see this, just recall that we have $E[X] = \frac{1}{2}$ for a continuous random variable $X \sim U(0, 1)$. So, all in all, we get

$$L_x = l_{x+1} + \frac{l_x - l_{x+1}}{2} = \frac{1}{2}(l_x + l_{x+1}).$$

Definition 5.6 (Life expectancy at age x). The *life expectancy at age x*, $x \in \{0, 1, ..., x_{max}\}$, is defined as

$$E_x := \frac{1}{l_x} \sum_{y=x}^{x_{max}} L_y.$$

Remark 5.7. The idea behind the formula given in the definition above is quite simple: The life expectancy of a person aged x is the number of years the survivors at age x live thereafter (that's the part with the sum) divided by the number of survivors l_x. So, E_x can be interpreted as the average number of years a person aged x has left to live thereafter.

Throughout the remaining part of the chapter, the term "life expectancy" is abbreviated by "LE". Now, we have two measures which provide information on the "expected remaining lifetime" we were initially interested in, as stated at the beginning of this chapter. Both are intuitively clear, but the question arises whether the underlying concepts are moreorless the same or not. This issue is of special importance when one wants to make group comparisons based on those quantities. For example, if you take the epilepsy dataset, one may think of calculating mean residual lifes for the patients and compare them with the corresponding life table life expectancies.

But, note that such a direct comparison is questionable due to the fact that we use two different formulas here. So, let's focus on the life table formula and, at first, re-write this quantity in the following proposition (Zimmermann [31]).

Proposition 5.8. *The following equality holds for all* $x = 0, 1, ..., x_{max}$:

$$E_x = \frac{1}{2} + \frac{1}{2} \prod_{j=x}^{x_{max}} (1 - q_j) + \sum_{k=x+1}^{x_{max}} \prod_{j=x}^{k-1} (1 - q_j).$$

Proof. According to the definitions of the life table quantities, we have

$$E_x = \frac{1}{l_x} \sum_{k=x}^{x_{max}} L_k = \frac{1}{2l_x} \sum_{k=x}^{x_{max}} (l_k + l_{k+1})$$

$$= \frac{1}{2l_x} \left(\sum_{k=x}^{x_{max}} l_k + \sum_{k=x+1}^{x_{max}+1} l_k \right)$$

$$= \frac{1}{2l_x} \left(l_x + l_{x_{max}+1} + 2 \sum_{k=x+1}^{x_{max}} l_k \right). \tag{5.2}$$

For every $k \geq 1$, the following line holds due to definition:

$$l_k = l_x \prod_{j=x}^{k-1} (1 - q_j), k \in \{x + 1, x + 2, ..., x_{max} + 1\}.$$

Now, to put things together, observe that l_x cancels out when applying this result to (5.2). So, finally, we have

$$E_x = \frac{1}{2} + \frac{1}{2} \prod_{j=x}^{x_{max}} (1 - q_j) + \sum_{k=x+1}^{x_{max}} \prod_{j=x}^{k-1} (1 - q_j).$$

$\square$

Remark 5.9. We have already stated in a previous remark that the life table quantities such as l_x only depend on the death probabilities. Now, we have actually proven this fact for E_x. Thus, the key message of the proposition above is that indeed, E_x only depends on the death probabilities $q_j, j = x, x + 1, ..., x_{max}$. In particular, we see that it does not matter which value the initial population size l_0 is set to because this number as well as all the other (hypothetical) quantities derived from it do not appear in the formula above.

Next, we write the result of Proposition 5.8 in a more probability theoretic way:

Proposition 5.10. *For all* $x \in \{0, 1, ..., x_{max}\}$, *we have*

$$E_x = \frac{1}{2} + \frac{S(x_{max} + 1)}{2S(x)} + \frac{1}{S(x)} \sum_{k=x+1}^{x_{max}} S(k).$$

Proof. Let T denote age at death. Note that, according to the definition of q_j, we have

$$P(T > j + 1 | T \geq j) = 1 - P(j \leq T < j + 1 | T \geq j) = 1 - q_j$$

for $j = 0, 1, ..., x_{max}$. If we now rewrite some terms appropriately, we get

$$1 - q_j = P(T > j + 1 | T \geq j) = \frac{P(T > j + 1)}{P(T \geq j)} = \frac{S(j + 1)}{S(j)},$$

$j = 0, 1, ..., x_{max}$. Putting this result together with Proposition 5.8 leads to

$$E_x = \frac{1}{2} + \prod_{j=x}^{x_{max}} \frac{S(j + 1)}{S(j)} + \sum_{k=x+1}^{x_{max}} \prod_{j=x}^{k-1} \frac{S(j + 1)}{S(j)}$$

$$= \frac{1}{2} + \frac{S(x_{max} + 1)}{2S(x)} + \frac{1}{S(x)} \sum_{k=x+1}^{x_{max}} S(k).$$

$\square$

Remark 5.11. Note that the part with the sum in Proposition 5.10 looks a bit like a "discretized" version of the residual life integral formula. However, the range of the summation index is not exactly the same as the integration interval. But maybe this "missing part" is the reason why there are two additional summands in the formula of Proposition 5.10. The factor $\frac{1}{2}$ somehow reminds me of the assumption that the deaths are uniformly distributed within the exact age intervals. So, maybe, we could as well derive the formula above using a random variable T for age at death which is made up by a weighted sum of uniformly distributed random variables, where the annual death probabilities are taken as weights.

Despite the uncertainities in interpreting the result of the proposition above, this statement points to a very useful fact: Obviously, all we need to calculate E_x is the survival function S! Thus, we can obtain life expectancies not only for the population of a whole country, but also for a study cohort since all we have to do is to calculate any "appropriate" estimate(s) of S and use the life table LE formula from above, then. At this point, it is worth mentioning that we could do it the other way round as well: Alternatively, we may use the mean residual life formula for the Weibull model and the life tables. For the latter, the integral would actually degenerate to a sum because the survival function for the population is a step function. Anyway, we have decided to use the life table LE formula, and so let's only examine this approach. However, before we illustrate these calculations by referring to our epilepsy dataset once again, we have to discuss an issue which is important when we want to compare patient and population LEs (sometimes, I will also call these quantities "model LEs" and "life table LEs", respectively). This will be the main topic of the next paragraph.

It seems quite natural to compare the LEs of a patient cohort to the LEs of some reference population. As several sorts of population lifetables are usually published online and can be downloaded for free, one may immediately think of using those tables for comparison purposes. For Austria, population lifetables are available that provide information on age-, gender-, year- and federal state-specific life expectancies (Statistik Austria [24]). Consequently, we are not restricted to some overall comparisons, but can account for several differences in the characteristics of the people. Thus, we can make comparisons for many subgroups (i.e., different covariate values) of the patient cohort. However, one should be cautious when directly calculating differences between life table and patient life expectancies: Recall that a population life table which is calculated, let's say, in 1980, only uses information of this particular year. More exactly speaking, the basic quantities in that life table are calculated based on the mortality experience of the population in 1980. As to the life expectancies, remember that only the annual mortality rates of 1980 are needed to get the LE values. Statistically speaking, we simply take the current mortality rates as estimates of the future ones.

Since we have decided to calculate the LEs for the patients using the same formula as proposed in the context of population life tables, it seems as if the desired comparisons can be made in a straightforward way. However, although the formula is the same, the underlying information is different: In contrast to the life tables, we usually follow at least some patients for more than one year, for example, from time of diagnosis to death or end of the study. Of course, if the study period is only one year, or, more generally speaking, "not too long" (e.g., 5 years), one need not worry about the "information discrepancy". But if the study period is fairly long, the problem mentioned above must not be neglected. In order to judge properly whether a given period is "too long", it is useful to have a look at the differences between the population life tables published at the start and the end of the study. For example, when examining the population lifetables of Tyrol between 1970 and 2010, you find that, roughly speaking, within 10 years, the life expectancies increase by 2 years (which points to a corresponding change in the annual death rates). Clearly, such a degree of change is not negligible.

Thus, we have to seek for a proper method to bridge that difference in the amount of information underlying population and patient LEs. Since we have life tables for various years at hand, I suggest using an algorithm that enables us to calculate "dynamic" life tables. This solution to the problem outlined above can be best explained by using our epilepsy data example. Consider a patient who was diagnosed at the age of 50 in 1980. Note that the type of epilepsy does not matter here. Likewise, *gender* is also omitted, although it is of course an important variable because we have gender-specific quantities in the life tables (and, of course, also in the model). But I suppose nobody would take death probabilities for females if asked to calculate life table quantities for a male person. Thus, as far as the description of the algorithm above is concerned, these two variables are omitted for sake of notational simplicity.

Once again, I want to point out that this person is not actually a patient who was recorded in our database, but represents a certain choice of covariate values we use in our fitted model to calculate survival probabilities and, finally, the corresponding LE value. In order to get a comparable value from the life tables, I propose the following procedure: As we have seen before, the LE formula involves the number of people alive at age x. Let us denote this quantity by $l_{x,y}$, where x and y denote age and year, respectively. Accordingly, $p_{x,y}$ is the survival probability for these input values (make sure that you don't confuse these "life table functions" with the ones defined above: The number of arguments they require is different, which, I hope, makes distinguishing between them easier). Now, the main goal is to calculate the numbers of people alive at age $50, 51, ..., 96$. Once we have these values, we are done (just remember the life table LE formula given in a previous section of this chapter). The first two quantities are $l_{50,1980}$ and $l_{51,1980}$. Next, we calculate the number of survivors at age 52: Recall that usually, this quantity is calculated by multiplying $l_{51,1980}$ with $p_{51,1980}$. But we want to account for the fact that due to the reasons stated above, we should better use the survival probability for 1981 instead of 1980. Thus, we get the desired number of survivors as

$$l_{52,1981} = l_{51,1980} \cdot p_{51,1981}.$$

Likewise, we continue with calculating $l_{53,1982} = l_{52,1981} \cdot p_{52,1982}$, and so on.

By the way, note that word "survivors" must not be taken literally: As already mentioned in Remark 5.5, the "number of survivors" contained in a life table does not refer to the number of people alive in any real population! If this was the case, that number could change due to other reasons than death, too (e.g., if some people move away). But, keep in mind that the number of "survivors" calculated in the life table is nothing but a hypothetical quantity which should facilitate interpretations, so we don't have to care about that issue: Actually, as already mentioned in Remark 5.9, all we need for life table calculations are the annual death probabilities.

Now, let us formulate the method suggested above in a more general way. Suppose a patient is diagnosed at age x in the year y, where x takes values from 0 to 94 and y from 1970 to 2009. The upper bounds of the ranges can be explained as follows: For a person aged 95, we only need the values l_{95} and l_{96} to calculate his or her life expectancy, so there is no need for looking at future life tables. Thus, the values we have in the tables are already the ones we need. Secondly, since our study ends in 2010, there's no need to calculate a "dynamic" life table for this year: If we used any information about the survival experience in the years after 2010, the results would again be biased.

Our goal is to calculate the life expectancy based on the life table values corresponding to the patient characteristics, which is equivalent to calculating the numbers of survivors for all ages from x to 96, which are denoted by $l_x, \ldots, l_{96}$. Now, according to the idea outlined above, we set up the following algorithm:

$$l_x := l_{x,y}, \quad l_{x+1} := l_{x+1,y}$$

$$l_{x+k} := \begin{cases} l_{x+k-1} \cdot p_{x+k-1,y+k-1} & k \in \{2,3,\ldots,\min(2011-y,96-x)\} \\ l_{x+k-1} \cdot p_{x+k-1,2010} & k \in \{2012-y,\ldots,96-x\} \end{cases}$$

The first line shouldn't be hard to understand. Then, the numbers of survivors are calculated according to the scheme outlined in the example above. In this context, some thoughts on the bounds of k are needed. To begin with, it is quite obvious that the values $2011 - y$ and $96 - x$ in the second line are due to the fact that $y + k - 1$ and $x + k$ must not exceed 2010 and 96, respectively. But additionally, we have to take the minimum of those two bounds since we want the algorithm to stop if k reaches at least one of them. Now, we have to distinguish between two cases:

(1) $96 - x \leq 2011 - y$ means that we have reached l_{96}, so we're done.

(2) If we have $96 - x > 2011 - y$, there are still some l_{x+k} values left to calculate. To illustrate that, let us turn back to our example once more: A person aged 50 in 1980 will be 80 years old in 2010. But we don't have any further life tables for subsequent years, so what should we do now? Well, at this point, we apply the "standard" method of life table construction and use the annual survival probabilities of the life table 2010 for the calculation of the remaining l_{x+k} values. Note that this corresponds to the situation we have in the study cohort: Of course, we get survival probabilities which cover a time period beyond 2010 from our Weibull model. But the data used to fit the model doesn't provide information on the survival experience of patients beyond 2010 since the study was terminated in that year. Thus, when calculating model LEs, we do extrapolation like in the algorithm above, with the only difference that we assume a certain trend (positive, constant or negative, depending on the shape parameter) of the survival probabilities in the model, whereas in the last line of the algorithm, we carry the status quo forward. Exactly speaking, for the latter part of the algorithm, we assume that the probability of dying at a certain age x stays the same for all the years after 2010.

Now, we know how to calculate the "right" life expectancies based on the life tables available, which then can be used for comparisons with the corresponding life expectancies of the study cohort. So, we are ready to look at the differences between patient and population LEs. However, keep in mind that at least for the patient LEs, we have to take the variance of the estimates into account.[2] Consequently, it is desirable to have confidence intervals for those LEs (as we assume the LEs we get from the life tables to be fixed numbers, we can simply subtract them from the confidence interval limits at the end). Exactly speaking, as we have discussed Weibull regression models in chapter 3, we want to stay with this setting and construct pointwise confidence intervals for a fixed set of covariates and a particular time point t.

To begin with, let's recall where the variance in the LE estimates originates from: Basically, the estimated parameters of the Weibull model come with some uncertainty, which causes the LEs to involve some variability, too. But how does this "translation" actually work? An answer to this question can be given by using the following well-known theorem.

Theorem 5.12 (Multivariate Delta Method). *Let* $(\mathbf{X_n})$ *be a sequence of* $d-$*dimensional random vectors,* $d \in \mathbb{N}$. *Furthermore, let* (a_n) *be a sequence in* $\mathbb{R}$ *such that* $\lim_{n \to \infty} a_n = \infty$. *Let* $\mathbf{x} \in \mathbb{R}^d$ *and* $\mathbf{Y}$ *be a* $d-$*dimensional random vector satisfying*

$$a_n(\mathbf{X_n} - \mathbf{x}) \xrightarrow{d} \mathbf{Y}.$$

Then, the following statement holds for any mapping $g \in \mathcal{C}^1(\mathbb{R}^d, \mathbb{R}^j)$:

$$a_n(g(\mathbf{X_n}) - g(\mathbf{x})) \xrightarrow{d} Dg(\mathbf{x})\mathbf{Y}.$$

Proof. See Van der Vaart [28], p. 25. $\square$

---

[2] Be aware of the fact that even the life table LEs are only estimates of the "true" LEs, see Statistik Austria [23], p. 24. But for sake of simplicity, we consider those point estimates as fixed values and take only the variability of the patient LEs into account.

Theorem 5.13 (Variance of model LE for the Weibull model). *Let us assume a Weibull PH model for survival time T. Moreover, let $\boldsymbol{\theta} := (\lambda, p, \beta_1, \beta_2, ..., \beta_k)'$ and $\hat{\boldsymbol{\Theta}}_n := (\lambda_n, p_n, \beta_{1,n}, \beta_{2,n}, ..., \beta_{k,n})'$ denote the vectors containing the model parameters and their ML estimates, respectively. Then, we have*

(i) *$\sqrt{n}(\hat{\boldsymbol{\Theta}}_n - \boldsymbol{\theta}) \xrightarrow{d} \mathbf{Y}$, where $\mathbf{Y} \sim \mathcal{N}_{k+2}(\mathbf{0}, \Sigma(\boldsymbol{\theta})^{-1})$.*

(ii) *If we consider a Weibull regression model and assume t and covariate values $x_1, x_2, ..., x_k$ to be fixed, the model LE (model LE estimator) only depends on $\boldsymbol{\theta}$ ($\hat{\boldsymbol{\Theta}}_n$).*

(iii) *Let $f(\hat{\boldsymbol{\Theta}}_n)$ denote the model LE estimator from (ii). Then, we have*

$$\sqrt{n}(f(\hat{\boldsymbol{\Theta}}_n) - f(\boldsymbol{\theta})) \xrightarrow{d} \mathcal{N}_1(0, grad(f)(\boldsymbol{\theta})'\Sigma(\boldsymbol{\theta})^{-1}grad(f)(\boldsymbol{\theta})).$$

(iv) *The model LE estimator $f(\hat{\boldsymbol{\Theta}}_n)$ has an asymptotic univariate normal distribution with mean $f(\boldsymbol{\theta})$ and variance $grad(f)(\hat{\boldsymbol{\Theta}}_n)'I(\hat{\boldsymbol{\Theta}}_n)^{-1}grad(f)(\hat{\boldsymbol{\Theta}}_n)$.*

Proof. For the proof of (i), we refer to Kalbfleisch and Prentice [14], pp. 179-181. Statement (ii) directly follows from Proposition 5.10. To see (iii), just apply the multivariate delta method (note that $f : (0, \infty)^2 \times \mathbb{R}^k \to \mathbb{R}_{\geq 0}$ is in $\mathcal{C}^1((0, \infty)^2 \times \mathbb{R}^k, \mathbb{R}_{\geq 0})$ due to Proposition 5.10 and the fact that the Weibull regression survival function is smooth with respect to the parameters).

To prove (iv), we need some consistency results for the Weibull MLEs, which can be found in Kalbfleisch and Prentice [14], p. 180. In particular, it can be shown that $\hat{\boldsymbol{\Theta}}_n$ is consistent for $\boldsymbol{\theta}$ and $\Sigma(\boldsymbol{\theta})$ can be estimated consistently by the observed Fisher information $n^{-1}I(\hat{\boldsymbol{\Theta}}_n)$. From the first result, we get $grad(f)(\hat{\boldsymbol{\Theta}}_n) \xrightarrow{P} grad(f)(\boldsymbol{\theta})$ by applying the continuous mapping theorem. So, all in all, the result from (iii) and Slutzky's Lemma (see Van der Vaart [28], pp. 7-8, 11, for the statements from asymptotic statistics mentioned in the text) yield

$$\sqrt{grad(f)(\hat{\boldsymbol{\Theta}}_n)'I(\hat{\boldsymbol{\Theta}}_n)^{-1}grad(f)(\hat{\boldsymbol{\Theta}}_n)}(f(\hat{\boldsymbol{\Theta}}_n) - f(\boldsymbol{\theta}))$$

$$= \sqrt{grad(f)(\hat{\boldsymbol{\Theta}}_\mathbf{n})'(n^{-1}I(\hat{\boldsymbol{\Theta}}_\mathbf{n}))^{-1}grad(f)(\hat{\boldsymbol{\Theta}}_\mathbf{n})}\sqrt{n}(f(\hat{\boldsymbol{\Theta}}_\mathbf{n}) - f(\boldsymbol{\theta}))$$

$$\xrightarrow{\text{d}} \sqrt{grad(f)(\boldsymbol{\theta})'\Sigma(\boldsymbol{\theta})^{-1}grad(f)(\boldsymbol{\theta})} \cdot \mathbf{Y},$$

where $Y \sim \mathcal{N}_1(0, grad(f)(\boldsymbol{\theta})'\Sigma(\boldsymbol{\theta})^{-1}grad(f)(\boldsymbol{\theta}))$. $\square$

Thus, we now know the approximate distribution of the LE quantity we are interested in, which enables us to calculate confidence intervals for life expectancies. From a practical point of view, these calculations can be quite time-consuming, especially when we want to obtain confidence intervals for a considerable amount of different covariate combinations. In addition to that, partial derivatives of the LE function are needed. Of course, this is not a serious problem, but forces us to implement the algorithm using different software packages, namely R and Mathematica (Wolfram Research, Inc. [30]), because in the latter one, formal operations like the derivations mentioned above are much easier to do than in statistical software. In the following example, we describe the main steps of such a confidence interval calculation procedure.

Example 5.14. At first, we have to pick a model which fits the data reasonably well. For sake of simplicity, we, again, assume a Weibull model for the data. Now, basically all we need are the maximum likelihood estimates and the estimated covariance matrix, which can be easily extracted from the model fit in R: Just use the functions survreg and ConvertWeibull, as already mentioned in previous examples. After having saved those estimates as some .txt or .csv file, we turn to Mathematica and basically carry out the calculations suggested by Theorem 5.13. More to the point, we calculate the partial derivatives of f evaluated at the maximum likelihood estimates (observe that maybe, we have to do this repeatedly because if we are interested in the life expectancies for different choices of the covariate vector $\mathbf{x}$, we, accordingly, have to consider different model LE functions f!). Due to Theorem 5.13, this yields the asymptotic variance of the model LE estimate. Then, we are ready to calculate an approximate $1 - \alpha$ confidence interval, which is saved in a .csv file. Finally, if we are interested in comparing model and life table LEs, we use R again and simply subtract the life expectancy based on the "dynamic" life table approach (recall that the life table LE is considered to be fixed!) from the lower and upper limit, respectively, of the model LE confidence interval.

We have already mentioned the fact that the method proposed above can be computationally exhaustive, e.g. when calculating confidence intervals for a fairly large number of covariate combinations. Nevertheless, the increase in time needed to do these calculations is not a very serious problem: Actually, you can get the results within a reasonable time span of approximately two days, so there is no need to worry about that issue.

However, I discovered two aspects which may be much more problematic: At first, I want to draw attention to the fact that the distributional result given in the theorem above is merely asymptotic. Of course, this for itself is not an issue one has to worry much about since asymptotic results are used for the construction of confidence intervals in a broad variety of statistical branches. But, we have used asymptotic results several times, as can be seen when having a closer look at the proof once more. For example, even the normality assumption for the model parameters, which was the point we started from, holds true only asymptotically. For example, "extreme" values of certain covariates could seriously affect the validity of the asymptotic distributional form assumed for the parameter estimators (Kalbfleisch and Prentice [14], p. 66). Likewise, there are some asymptotic results needed to prove the delta method itself.

Secondly, we have to take into consideration that in certain settings, such as in the examination of our epilepsy dataset, further refinements of the LE calculations are needed. Exactly speaking, there are substantial arguments against the assumption that the survival (or death) probabilities of the patients follow the Weibull model over the entire life span of the subjects. At first sight, this seems to be a question of modelling rather than an issue concerning the reliability of the results we get by using the delta method. But, as the reader will see in the modified formula stated below, the "revised" LE function is (probably) not differentiable. So, applying the delta method wouldn't be possible any more. Therefore, we have to seek for another method to calculate confidence intervals. But, before we do that, let's turn back to the modification of the LE formula mentioned above. The main idea is best illustrated by using our epilepsy dataset, although one should keep in mind that the underlying problem could arise in many similar research questions.

Let's recall that we inherently assume in our Weibull model that the risk of dying (i.e., the hazard function) is either increasing, constant or decreasing as we proceed on the timeline, starting at time of diagnosis. However, this assumption is highly questionable: For example, the Austrian period life table of 2000/02 shows that starting at age 30, the death probabilities increase monotonically. When looking at our estimate of the shape parameter p, we find that it is greater than 1. So, the model death probabilities also increase monotonically. But, no matter how small or large the estimate of p actually is, as long as it is above 1, the model death probability function, say, $p(t)$, which is given as

$$p(t) := P(t \leq T < t+1 | T \geq t) = 1 - \frac{S(t+1)}{S(t)} = 1 - exp(\lambda(t^p - (t+1)^p)),$$

is dominated asymptotically by t^c for **any** $c \in (0,1)$ (i.e., $p(t) = o(t^c)$, $c \in (0,1)$, for $t \to \infty$). By contrast, the increase seen in the corresponding probabilities from the life tables is of order $o(t^k)$ for some fixed $k \geq 1$ (by the way, this behaviour can be observed in the regional life tables of Tyrol, too). Thus, at a "sufficiently large" time point from diagnosis, the model probabilities will be fairly unrealistic. As a result, the model life expectancies will be biased, especially for quite small values of the covariate *age at diagnosis*.

Basically, there are two ways of dealing with this problem. Firstly, we could pick another model which is more flexible than the Weibull model. For example, the Cox proportional hazards model allows for a broad variety of different shapes since the baseline hazard is unspecified and therefore estimated nonparametrically. Another option would be the generalized gamma model, which, for example, also allows for so-called "bathtub-shaped" hazards (see Cox et al. [7]).

Secondly, we could also specify some "breakpoint" value t_b that defines the time point after diagnosis from which on we take, for example, the maximum of the model and the life table death probability. This modification can be interpreted as follows: From that certain time point t_b after diagnosis onwards, we assume that the patients' risk of dying is greater or equal than the risk of dying for the reference population. In my opinion, such an assumption makes sense because even if you take protective effects (i.e., the patients are scheduled for regular medical checks, in contrast to the majority of the reference population) into account, it would be very unlikely that the discrepancy between the patients and the reference population could be as high as indicated by the extremely different orders of increase mentioned above.

As to our concrete epilepsy dataset, we now examine the two options of solving the problem.

Example 5.15. First, we fit a Cox model and a Weibull regression model with the same covariates as we used in chapter 3 and have a look at the (cumulative) baseline hazard (see Figure 5.1) as well as the difference in LEs based on the two models for a certain choice of covariate and t values. Both analyses show hardly any discrepancies, except a slight change in the hazard's shape at 20-25 years following diagnosis insofar as the increase seems to slow down a bit. But as far as the baseline hazard comparison is concerned, note that for calculating an estimate of the Cox model baseline cumulative hazard, the R function **basehaz** contained in the **survival** package uses the mean

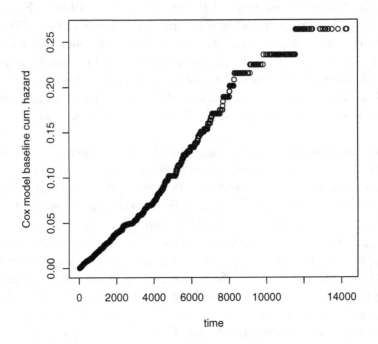

Figure 5.1: Baseline cumulative hazard estimate for a Cox PH model
calculated with the function `basehaz`

values of each covariate. Apparently, this does not make any sense
for nominal variables. Nevertheless, the result of `basehaz` provides at
least some general idea of the hazard's outlook, which should do for
the moment.
In addition to that, a spline-based approximation of the hazard for
each of the four age groups <25, 25-45, 46-65 and >65, without taking any further covariates into account, is calculated by using the R
function `heft` included in the package **polspline** (Kooperberg [17]).
Interestingly, the plots for the first two groups show a hazard function

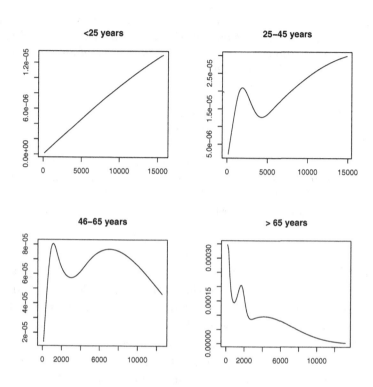

Figure 5.2: Spline-based approximation of the age-group-specific hazards

which increases, whereas the hazards for the latter two groups decrease towards the end (Figure 5.2).

By the way, at first sight, this result contradicts the proportional hazards assumption. But remember that we haven't taken any other covariates than *age at diagnosis* into account. So, to judge if the Weibull (or Cox) model assumed for the data is indeed inappropriate, we fit a hazard regression model which is based on splines to the data, allowing for the covariates we have included in the Weibull model as well as for various interactions between them, which are

Table 5.1: Results of `hare` function call for the epilepsy data. Note that one line of the R output which contains only some information about a certain knot value for *year of diagnosis* was omitted in this table.

	coefficient estimate	standard error	Wald statistic
intercept	-120	49	-2.54
gender	100	34	2.94
age at diagnosis	0.079	0.0039	20.07
year of diagnosis	0.056	0.025	2.26
gender × year of diagnosis	-0.05	0.017	-2.93

examined by the R function **hare**, which is also included in **polspline**, "automatically". The results of the fit are displayed in Table 5.1.

The set of variables which are finally included is very similar to the one we have used for the Weibull model (see Example 3.7), apart from the fact that the two dummy variables coding the type of epilepsy were excluded during the modelling procedure in R. However, these coefficients are not significant in the Weibull AFT regression model, too, so this issue won't have serious consequences. Moreover, one additional interaction term, namely between *gender* and *year of diagnosis*, is incorporated in the hazard regression model. Based on the Wald statistic, we compute the corresponding p-value and find that it is about 0.003. So, indeed, this term turns out to be significant. However, for medical reasons, *etiology* should be accounted for. Of course, adding the two dummies to the model could affect the significance of the interaction between *gender* and *year of diagnosis*. So, summing up, we have found some evidence that the Weibull model specified in previous chapters might be improved by adding this additional term. However, the method underlying the **hare** function is just one possible approach among others dealing with model building. Therefore, as most of the results are consistent with our previous findings, we most likely don't make a serious mistake if we stay with our model suggested in previous chapters, although the **hare** procedure is certainly worth having a closer look at.

To turn to the second way of improving the LE values, there may be some need for a modification concerning the use of the model death probabilities for calculating life expectancies: In general, as stated before, we expect the death probabilities to start increasing after some time has passed since diagnosis. But, obviously, such a trend cannot be seen in the data. Somehow, that's not very surprising: Although the maximum observation period is 40 years, we should keep in mind that the majority of the patients was diagnosed in the 2000's, so the observation period may be less than 10 years in most cases. Even the nonparametric hazard estimates calculated for the four age groups mentioned above can at most be interpreted as a slight indication of the death probability pattern we expect: Remember that these hazard estimates might be subject to confounding since we do not adjust for any other covariates. Of course, we could generate plenty of subgroups according to the large amount of possible covariate-combinations and calculate separate hazard estimates for them. But as already mentioned in the model checking chapter, we would have to face serious problems due to very small sample sizes.

Thus, we have to seek for a meaningful breakpoint value by other means than (non-)parametric modelling of the survival data. For example, we could calculate *Standardized Mortality Ratios (SMRs)*, which represent, loosely speaking, the number of observed deaths divided by the number of expected deaths, that is, the number of deaths if the death probabilities of the reference population applied to the study cohort. Especially for the age groups $0 - 24$ and $25 - 45$, the SMRs for subsequent years following diagnosis decrease at first, but slightly increase towards the end The point where the trend changes is at approximately $t_b = 20$ years after diagnosis (Trinka et al.[27]). Based on these results, I suggest using the following modified procedure for calculating life expectancies: For the age groups $46 - 65$ and $65 - 75$, we use the Weibull lifetable LE formula as defined above. For the age groups $0-24$ and $25-45$, we stay with our Weibull model for the time points $0, 1, ..., 20$ years after diagnosis, but replace the Weibull death probabilities by the maximum of the Weibull and the lifetable death probabilities for all time points beyond $t_b + 1 = 21$ years. Note that some caution is needed here: We must not take the probabilities from the life table of the diagnosis year only, but follow some "dynamic" idea similar to the concept behind the "dynamic" life expectancies described above. For example, if we want to calculate the "modified" death probability at $t = 25$ for a patient who was diagnosed at an age of 25 in 1980, we have to calculate the maximum of the corresponding Weibull quantity and the death probability of a 50-year-old in 2005(!). As our study was terminated at the end of 2010, we assume the death probabilities of the 2010 lifetable are reasonable estimates of the death probabilities in subsequent years.

Remark 5.16. By the way, we see an important aspect in the example above: Recall that for LE calculations, we often need to define a certain time point t_{max} which represents the maximum survival time, since an infinitely large value of t is not realistic in many applications. In the epilepsy dataset analysis, we decided to set this upper limit to $t_{max} = 95-$ *age at diagnosis*, which means that we assume nobody lives beyond the age of 95, to ensure that our model meets the assumptions of the period life table methodology as close as possible. To cut it short, all we have to do then is to calculate annual Weibull death probabilities up to an age of 95 (or, equivalently, up to $t_{max} = 95-$ *age at diagnosis*). So, for example, if you take a 40-year-old patient, you calculate death probabilities for $0, 1, ..., 55$ years after diagnosis. But, now remember that the data underlying our model was collected from 1970 to 2010. So, even if in our original dataset, there was a 40-year-old guy who was diagnosed in 1970 and still alive in 2010, he would at most be observed for 41 years. So, at least the last fourteen death probabilities are definitely extrapolations. We've touched this question already in the context of the "dynamic" life table calculations, where we said that we carry the status quo of 2010 forward to all subsequent years. So, summing up, the key message is that survival analytic models are not superior to the life table methodology as far as extrapolation is concerned. Secondly, one might criticize the modified LE calculation prodecure proposed in the example above, claiming that it is a somehow strange idea to fit a certain model to the data and seek for arguments supporting the underlying assumptions at first, but, then, refuse to "trust" the model for the whole time span. But, to explain this seeming contradiction, one should be aware of the fact that although it's a very important issue to choose a model which fits to the data "well", we can't get completely rid of the problem of making moreorless reasonable assumptions about the death probabilities that exceed our original observation range. In other words, even if we decided to take the model probabilities for the entire time span, we would not be able to judge by looking merely at the patients' survival data whether **each** of the model values as well as the underlying assumptions concerning trends etc. are plausible or not.

After this relatively extensive account of the epilepsy dataset, let's recall that the delta method may be unreliable at least in some cases. In addition to that, it cannot be applied to our modified life expectancy calculations. Or, speaking more generally, what if the LE formula we decide to use is not a function contained in $\mathcal{C}^1(\mathbb{R}^d, \mathbb{R}^j)$? Then, we could turn our focus on a bulk of several resampling methods that are referred to by the term *bootstrapping*. Of course, I cannot give an extensive overview here, but I will make some general remarks on this issue first and then describe the bootstrap method I have chosen.

To consider a more general setting than just the LE confidence interval problem, let us assume that we have a random sample $\mathbf{X}$ from a certain population which is used to calculate an estimator $\hat{\theta}(\mathbf{X})$ of an unknown population parameter θ. Now, if you want to make inference about θ and the distribution of $\hat{\theta}(\mathbf{X})$ is known, you can proceed in a straightforward way. However, in many cases, it may be hard - if not impossible - to determine the (exact or asymptotic) distribution of the estimator. Then, one can make use of the following idea called the *bootstrap principle*, which basically means that we we replace θ (the unknown parameter) by $\hat{\theta}(\mathbf{x})$ (the estimate calculated based on a sample $\mathbf{x}$). Then, we take a resample $\mathbf{x}^*$, the so-called *bootstrap sample*, based on $\mathbf{x}$ (I will explain the details in a second) and calculate $\hat{\theta}(\mathbf{x}^*)$. We repeat this resampling procedure B times, where B is a "sufficiently large" positive integer. Of course, the "substitution" in this procedure usually comes at the price of some error. But the idea is that by resampling (or, equivalently speaking, simulating data from our sample which we assume to represent the population), we eventually get "reliable" information about the distribution of $\hat{\theta}(\mathbf{X})$. By "reliable", I mean that using this method, we might be able at least to compensate for the error mentioned above, if not, ideally, improve the results of other, classical procedures, for example, as far as the coverage of confidence intervals is concerned.

There are many different resampling plans available in statistical literature. Usually, they are divided into two main groups: By the term *nonparametric bootstrap*, we address sampling with replacement from the vector $\mathbf{x}$. If we assume a parametric model $P = \{P_\theta : \theta \in \Theta\}$ underlying the data (e.g., a linear model), the bootstrap samples are taken from $P_{\hat{\theta}(\mathbf{x})}$. This procedure is called *parametric bootstrap*.[3] Each of these families comprises various methods, but for now, I only describe the method I would suggest for calculating LE confidence intervals:

1. **Estimation**: To begin with, we calculate an estimate $\widehat{LE}_{t,\mathbf{x}}(X)$ of the desired quantity $LE_{t,\mathbf{x}}$ (life expectancy at time t after diagnosis for a subject with covariate values $\mathbf{x}$) from the sample X, where X denotes a matrix whose columns contain the survival times, the censoring indicators and the covariate values (note that X must not be confused with a random vector, which would be printed boldly). Especially when we assume, for example, a

---

[3] Especially in regression contexts, also a so-called *semiparametric bootstrap* can be used, see Carpenter and Bithell [3].

Weibull model for the data, the phrase "calculating an estimate of LE" actually means that we fit a Weibull model to the data and calculate the desired estimate afterwards.

2. **Resampling**: We follow a procedure called *case resampling*, which is a method of nonparametric resampling and means that we simply draw samples $\mathbf{x_i^*} := (t_i, \delta_i, \mathbf{x_i}), i = 1, 2, ..., n$, with replacement from the rows of our original data matrix X, where n is equal to the number of rows of X, i.e., the number of subjects enrolled in the study. If we denote the resample matrix by X^*, we now calculate $\widehat{LE}_{t,\mathbf{x}}(X^*)$. We repeat this process $B = 2000$ times, which is a number that should work pretty well (see Carpenter and Bithell [3]). So, we get a vector

$$\widehat{\mathbf{LE}}^* := (\widehat{LE}^{(1)}, \widehat{LE}^{(2)}, ..., \widehat{LE}^{(2000)})$$

containing the LE estimates based on the bootstrap samples, where $\widehat{LE}^{(k)}$ denotes the estimate calculated in the k-th iteration step, $k = 1, 2, ..., B$.

3. **CI calculation**: Finally, we have to decide which method to take in order to actually build the desired confidence intervals. In the context of the resampling plan we would propose, a quite simple and popular choice is the so-called *percentile method*. Following this method, you just take the 0.025 and 0.975 empirical percentiles of $\widehat{\mathbf{LE}}^*$ as the left and right endpoints, respectively, of the confidence interval. Note that, however, one has to be careful when calculating confidence intervals in such a way because due to the underlying assumptions, the distribution of $\widehat{LE}_{t,\mathbf{x}}(X)$ must be symmetric. If this is not the case, we might get quite unsatisfactory results in terms of coverage. In addition to that, just as in linear regression, if we want to calculate confidence intervals for several time point-covariate-combinations, keep in mind that we have to adjust the α-level.

It should be noted that the first problem mentioned above can be circumvented by using a more advanced calculation procedure, namely the so-called *bias-corrected and accelerated bootstrap (BCa) method*. In R, you simply apply the function `bootBCa` contained in the **rms** package (Harrell [12]) to the results from steps 1 and 2 to get the desired confidence interval, which has better coverage properties than the simple percentile interval (see Carpenter and Bithell [3], where a theoretical account of this method can be found).

Epilogue

Instead of merely summarizing the main results of the previous chapters, I want to give a short outline of the "pathway" which led to my master's thesis. In this short final part, the focus won't be on scientific writing any more. Now, the main goal is to give the reader the possibility to see how I actually worked on this topic.

As mentioned in the preface, the whole story began some weeks before Christmas in 2013. At that time, I had just started with the master's programme. I was already interested in probability theory and statistics, but I had never been asked to analyze a real-life dataset before. Therefore, I wasn't completely sure if I would be able to do the epilepsy data analysis. However, as mentioned before, I agreed to join the research team, and after having completed my courses in February 2014, I started working on the project. At first, I conducted the data import in R. Everything looked nice, and so, I was a bit surprised when I received an email from Professor Bathke, who asked me if I had problems with the dates contained in the database. Then, I tried to do some calculations with the dates and found out that things weren't that easy as they seemed to be at first sight

To cut it short, in the weeks after this first error, I got more and more used to working with R. So, I turned my focus on methodological rather than technical questions. At first, I studied the basics of regression modeling and conducted some checks of the underlying assumptions, as discussed in chapter 3 and 4. Then, I proceeded with thinking about the main topic of both my master's thesis (chapter 5) and the medical research project, namely, the calculation and comparison of life expectancies. To be precise, my colleagues wanted me to carry out an analysis similar to the one described in Gaitatzis et al. [8]. It took some time to figure out the details of this method, but by the end of May 2014, I finished the implementation of the algorithms in R and started calculating the point estimates of the differences in life expectancy between the study cohort and the population of Tyrol. Everything worked well, and so I met with Professor Trinka, Dr. Höller and Dr. Granbichler in June 2014 to discuss the results with them. They were very impressed, but suggested some modifications of the method I've chosen. In particular, Gaitatzis et al.[8] used the maximum of the model and the population death probabilities for the entire time range (this issue was mentioned in chapter 5). Professor Trinka and Dr. Granbichler argued that by doing so, any possible protective effects were wiped out (see chapter 5). Therefore, I re-calculated the life expectancy differences - and was shocked: When taking only the model death probabilities, the LEs of the patients exceeded those of the population by about 15 or even 20 years!

I checked the R code as well as the mathematical justification of my method several times, but I couldn't find any mistake. So, I carefully examined the appropriateness of the model - and, thus, got further insights into model checking and building, as discussed in chapter 4. I thought, for example, that maybe, there are some data points which seriously affect the model estimates. However, after having spent several weeks on thinking about this problem without finding any substantial argument against the model, I remembered what Dr. Granbichler had said about the extreme life expectancy differences: If we can be quite sure that our method is correct, we should try to find an explanation for the surprising results.

As I'm not a medical expert, I left this task to my research colleagues and started thinking about another methodological improvement: Gaitatzis et al.[8] didn't report any confidence intervals for the estimates of the life expectancy differences. I had a look at the re-written life table LE formula (chapter 5) and tried to figure out its relation to the model parameters. Suddenly, I remembered that in the statistics class, we had shortly discussed the multivariate delta method, which in this case appears to be exactly the right tool to calculate confidence intervals for the LE differences. The next time I met with Professor Bathke, I told him about my idea. As he said I should have a try, I implemented the procedures I had in my mind in R and Mathematica, which was a quite challenging and time-consuming task. Anyway, in November 2014 I was ready to submit the results to my research colleagues. We all were very satisfied with this substantial methodological improvement. But after some time, I was asked by Professor Bathke to present my method in his PhD students class. So, I set up a short presentation about the basic research question and, in particular, my confidence interval calculation method. After the presentation, Professor Bathke and Professor Trutschnig told me that basically, my derivations are correct. But they said I should be cautious when applying the method I proposed, because there may be problems in terms of coverage (see chapter 5). Therefore, I decided to carry out some simulations in R and Mathematica to examine this issue. Indeed, at least in some cases, the coverage was only around 90 per cent.

Moreover, I wasn't completely satisfied with another aspect of the method developed so far. Apart from the coverage problem, the LE differences were quite large, as mentioned before, and, to be honest, I couldn't believe that this was only due to some protective effect. However, one day, I suddenly realized what could be the main reason for the extreme LE differences: Maybe, we should not "trust" the model death probabilities for the entire time range, since this would be a too optimistic assumption! Indeed, after introducing a breakpoint value as described in chapter 5, the differences decreased substantially. As a result, however, the delta method couldn't be used any more, because the life expectancy function wasn't continuous. In other words, the improvement concerning the LE differences gave rise to another difficulty. Following the suggestions of Professor Pauly at a workshop in Ulm at the end of February 2015, I decided to replace the delta method confidence intervals by a bootstrap approach (see chapter 5).

The final step of the analysis was made in April 2015. At that time, I had already started writing down some parts of my master's thesis and, therefore, went through the theory behind population life tables once more. When thinking carefully about the underlying data generating process, I discovered a discrepancy between the study cohort and the population. This finally led to the idea of the so-called "dynamic" life tables discussed in chapter 5.

Summing up, one and a half years have passed now since I've started working on the epilepsy data analysis. Looking back, I'm glad that I took the chance to get in touch with applied statistics. Of course, there have been some drawbacks on the way. However, I think that every mistake and every suggestion from Professor Bathke and my research colleagues has eventually led to an important improvement of the statistical methodology. Although I sometimes really despaired of making hardly any progress for several weeks, I have thus learned what doing mathematical research is like. Moreover, I think such an analysis is definitely worth spending careful methodological thoughts on, because the data I was working with is not a toy dataset: Maybe, the results of such an analysis will influence medical or political decisions. I feel that the method I've proposed is not perfect, but I think it can serve as a good basis for further thoughts on this topic.

Bibliography

[1] O. AALEN. Nonparametric Inference for a Family of Counting Processes. *Ann.Statist.* **6**(4), 701–726, 1978.

[2] N. BRESLOW. Covariance Analysis of Censored Survival Data. *Biometrics* **30**(1), 89–99, 1974.

[3] J. CARPENTER AND J. BITHELL. Bootstrap Confidence Intervals: When, Which, What? A Practical Guide for Medical Statisticians. *Statist. Med.* **19**, 1141–1164, 2000.

[4] J. COCATO, P. PEDUZZI, T.R. HOLFORD AND A.R. FEINSTEIN. Importance of Events per Independent Variable in Proportional Hazards Regression Analysis I. Background, Goals and General Strategy. *J Clin Epidemiol* **48**, 1495–1501, 1995.

[5] D. COLLETT. *Modelling Survival Data in Medical Research.* Second Edition, London 2003.

[6] D.R. COX. Regression Models and Life-Tables. *J. Roy. Statist. Soc. Ser. B* **34**(2), 187–220, 1972.

[7] C. COX, H. CHU, M.F. SCHNEIDER AND A. MUNOZ. Parametric Survival Analysis and Taxonomy of Hazard Functions for the Generalized Gamma Distribution. *Statist. Med.* **26**, 4352-4374, 2000.

[8] A. GAITATZIS, A.L. JOHNSON, D.W. CHADWICK, S.D. SHORVON AND J.W. SANDER. Life Expectancy in People with Newly Diagnosed Epilepsy. *Brain* **127**, 2427–2432, 2004.

[9] P.M. GRAMBSCH AND T.M. THERNEAU. Proportional Hazards Tests and Diagnostics Based on Weighted Residuals. *Biometrika* **81**(3), 515–526, 1994.

[10] C.A. GRANBICHLER, W. OBERAIGNER, G. KUCHUKHIDZE, G. BAUER, J.P. NDAYISABA, K. SEPPI AND E. TRINKA. Cause-Specific Mortality in Adult Epilepsy Patients from Tyrol, Austria: Hospital-Based Study. *J Neurol* **262**(1), 126-133, 2015.

[11] A. HANIKA AND H. TRIMMEL. Sterbetafel 2000/02 für Österreich. *Stat Nachr Osterr Stat Zent Amt* **2**, 121–131, 2005.

[12] F.E. HARRELL JR. rms: Regression Modeling Strategies. R package version 4.3-1. URL http://CRAN.R-project.org/package=rms, 2015.

[13] S. HUBEAUX AND K. RUFIBACH. SurvRegCensCov. Weibull Regression for a Right-Censored Endpoint with Interval-Censored Covariate. R package version 1.3. URL http://CRAN.R-project.org/package=SurvRegCensCov, 2014.

[14] J.D. KALBFLEISCH AND R.L. PRENTICE. *The Statistical Analysis of Failure Time Data*. Second Edition, Hoboken 2002.

[15] E.L. KAPLAN AND P. MEIER. Nonparametric Estimation from Incomplete Observations. *J.Amer.Statist.Assoc.* **53**, 457–481, 1958.

[16] J.P. KLEIN AND M.L. MOESCHBERGER. *Survival Analysis: Techniques for Censored and Truncated Data*. Second Edition, New York 2003.

[17] C. KOOPERBERG. polspline. Polynomial Spline Routines. R package version 1.1.9. URL http://CRAN.R-project.org/package=polspline, 2013.

[18] W. NELSON. Theory and Applications of Hazard Plotting for Censored Failure Data. *Technometrics* **14**(2), 945–966, 1972.

[19] W. OBERAIGNER AND W. STÜHLINGER. Record Linkage in the Cancer Registry of Tyrol, Austria. *Methods Inf Med* **44**, 1–5, 2005.

[20] A.B. OWEN. *Empirical Likelihood.* New York, 2001.

[21] P. PEDUZZI, J. COCATO, A.R. FEINSTEIN AND T.R. HOLFORD. Importance of Events per Independent Variable in Proportional Hazards Regression Analysis II. Accuracy and Precision of Regression Estimates. *J Clin Epidemiol* **48**, 1503-1510, 1995.

[22] R CORE TEAM. R. A Language and Environment for Statistical Computing. R Foundation for Statistical Computing, Vienna, Austria. URL http://www.R-project.org/, 2014.

[23] STATISTIK AUSTRIA. Demographische Indikatoren 2012. Querschnittsindikatoren 2012. Schnellbericht 8.3. URL http://www.statistik.at, Wien 2014.

[24] STATISTIK AUSTRIA. Sterbetafeln Österreich. URL http://www.statistik.at, 12.05.2015.

[25] L. SUN AND Z. ZHANG. A Class of Transformed Mean Residual Life Models With Censored Survival Data. *J. Amer. Statist. Assoc.* **104**, 803–815, 2009.

[26] T. THERNEAU. A Package for Survival Analysis in S. URL http://CRAN.R-project.org/package=survival, 2014.

[27] E. TRINKA, G. BAUER, W. OBERAIGNER, J.P. NDAYISABA, K. SEPPI, C.A. GRANBICHLER. Cause-Specific Mortality among Patients with Epilepsy: Results from a 30-Year Cohort Study. *Epilepsia* **54**(3), 495–501, 2013.

[28] A.W. VAN DER VAART. *Asymptotic Statistics.* 8th printing, New York 2007.

[29] W. WEIBULL. A Statistical Distribution Function of Wide Applicability. *J Appl Mech* **18**, 293–297, 1951.

[30] WOLFRAM RESEARCH, INC. Mathematica. Version 8.0. Champaign, IL, 2010.

[31] G. ZIMMERMANN. Life Expectancy Comparison between a Study Cohort and a Reference Population. *Austrian Journal of Statistics* **45**(2), 53–67, 2016.

Printed in the United States
By Bookmasters